Remote Sensing and Image Interpretation

Remote Sensing and Image Interpretation

Editor

Beniamino Cipriani

scitus
academics

Remote Sensing and Image Interpretation
Edited by **Beniamino Cipriani**

ISBN: 978-1-68117-170-8
Library of Congress Control Number: 2015951994

SCITUS Academics LLC,
www.scitusacademics.com
Box No. 4766, 616 Corporate Way,
Suite 2, Valley Cottage,
NY 10989

Printed in United States of America on Acid Free Paper

Preface

Remote sensing and its related technologies, such as geographical information systems (GIS) and the Global Positioning System (GPS), have a widespread effect on science, government and business. Rapid progress in computer power and sensor design allows remote sensing and its allied technologies, such as geographical information systems (GIS) and the Global Positioning System (GPS), to play an extremely important role in science, engineering, resource management, trade and other areas of human activity.

Computer-based evaluation of remote-sensing image data is increasingly important as the types of digital image data from different aircraft and satellite sensors are increasing. The effective use of these remote sensing telemetry data involves the accurate delivery of the information contained in these data into applicable terms. And all these applications are growing very quickly, that include everything from various applications for land use and land cover mapping to the detection and monitoring of air pollution and oil spills, and earth surfaces and atmospheric measures. This book offers a comprehensive introduction to both the field of remote sensing and image analysis from latest developments in the digital image retrieval to the future age of satellite technology. This book is likely intended for advanced graduate students in remote sensing and image interpretation and also as a reference for the growing number of geospatial information practitioners.

Table of Contents

CHAPTER 1 **Unmanned Aerial Systems for Photogrammetry and Remote Sensing: A Review 1**

ABSTRACT 1
INTRODUCTION 2
On names and acronyms 2
Pioneers 3
General evolution 3
Literature evolution 4
Organization of the article 6
EARLY DEVELOPMENTS 6
UNMANNED AERIAL SYSTEMS AND UNMANNED AERIAL SYSTEMS FOR PARS 7
Unmanned aircraft 9
Ground control station 10
Communication 10
Mission planning 11
REGULATORY BODIES AND REGULATIONS 12
NAVIGATION, ORIENTATION AND SENSING PAYLOADS 16
Autopilots and navigation systems 17
Orientation systems 19
Sensing payloads 22
PROCESSING 28
Image orientation and camera calibration 29
Surface reconstruction 33
UAS PaRS APPLICATIONS AND GEOMATIC MARKETS 36

Agricultural and environmental applications....37
Aerial monitoring in engineering....39
Cultural heritage....40
Traditional surveying, conventional mapping and photogrammetry, and cadastral applications....41
CONCLUSIONS....42
APPENDIX A. UAS PICTURES AND COMMON PARS UAS....43
REFERENCES....45
CITATION....57

CHAPTER 2 A Review of Ocean Color Remote Sensing Methods and Statistical Techniques for the Detection, Mapping and Analysis of Phytoplankton Blooms in Coastal 59

ABSTRACT....59
INTRODUCTION....60
ALGAL BLOOMS IN THE CONTEXT OF THIS REVIEW....67
OCEAN COLOR REMOTE SENSING ALGORITHMS....69
Reflectance classification algorithms....70
Reflectance band-ratio algorithms....72
Bio-optical models....79
THE DETECTION OF SPECIFIC TYPES OF ALGAL BLOOMS....80
Algal blooms with surface expressions....80
STATISTICAL TECHNIQUES AND DATA ASSIMILATION TO ASSESS PHYTOPLANKTON BLOOM DYNAMICS....84
Statistical partitioning of marine ecosystems....84
Time-series, fitted models and signal processing techniques....86
Satellite product climatologies and merging data from multiple sources....87
CONCLUSIONS AND FUTURE DIRECTIONS....92
ACKNOWLEDGMENTS....94
REFERENCES....95
CITATION....117

CHAPTER 3 **Land Resources Assessment of El-Galaba Basin, South Egypt for the Potentiality of Agriculture Expansion Using Remote Sensing and Gis Techniques 119**

ABSTRACT.......... 119
INTRODUCTION.......... 120
MATERIAL AND METHODS.......... 121
Study area 121
Digital image processing.......... 123
Digital terrain analyses 125
Mapping units extraction.......... 125
Field survey 126
Soil analyses.......... 126
Land capability modeling.......... 126
Cropping pattern modeling.......... 128
RESULTS 129
Landscape characteristics 129
Mapping units of El-Galaba basin 131
Land use capability.......... 135
Cropping pattern 135
Cropping pattern 137
CONCLUSION 139
ACKNOWLEDGMENTS 139
REFERENCES.......... 139
CITATION.......... 142

CHAPTER 4 **Machine Learning in Geosciences and Remote Sensing 143**

Highlights.......... 143
ABSTRACT.......... 143
INTRODUCTION.......... 144
OVERVIEW OF ML APPLICATIONS IN GEOSCIENCES AND REMOTE SENSING.......... 145
ILLUSTRATIVE EXAMPLES.......... 149
Characterizing airborne particulates.......... 149
Dust sources.......... 153
CONCLUSION AND FUTURE DIRECTION 158

REFERENCES ... 159
CITATION ... 165

CHAPTER 5 Geophysical, Remote Sensing, Gis, and Isotopic Applications for a Better Understanding of the Structural Controls on Groundwater Flow in the Mojave Desert, California ... 167

HIGHLIGHTS ... 167
ABSTRACT ... 168
INTRODUCTION ... 168
GEOLOGIC AND HYDROGEOLOGIC SETTING ... 171
METHODOLOGY ... 174
Remote sensing methods ... 175
Geophysical methods ... 175
Isotopic methods ... 177
Geographic Information System (GIS) analyses ... 178
DATA AND FINDINGS ... 179
VLF and magnetic data ... 179
VES data ... 185
Stable isotope data ... 188
RESULTS AND REGIONAL IMPLICATIONS ... 193
ACKNOWLEDGEMENTS ... 195
REFERENCES ... 196
CITATION ... 199

CHAPTER 6 An Overview of Integrated Remote Sensing and Gis for Groundwater Mapping in Egypt ... 201

ABSTRACT ... 201
INTRODUCTION ... 201
WATER RESOURCES IN EGYPT ... 202
Elements of a groundwater map ... 207
1 - Hydrogeological units map ... 207
2 - Lithological map ... 208
3 - Surface water map ... 208
4 - Man made features map ... 208

5 - Groundwater features 208
6 - Topographic map 209
7 - Base of Aquifer map 209
8 - Isopiezometric map 209
9 - Isosalinity map 210
RS AND GIS FOR GROUNDWATER MAPPING 210
Applications of RS and GIS in groundwater mapping 210
GROUNDWATER MAPPING IN EGYPT USING RS AND GIS (CASE STUDIES) 217
Eastern Desert 217
Sinai Peninsula 222
Western Desert 224
Nile Delta and Nile Valley 226
CONCLUDING REMARKS AND RECOMMENDATIONS 227
ACKNOWLEDGMENTS 228
REFERENCES 228
CITATION 232

CHAPTER 7 Satellite Remote Sensing Outputs of the Certain Glaciers on the Territory of East Georgia 233

ABSTRACT 233
INTRODUCTION 234
RESEARCH AREA & GLACIERS SURVEY 235
RESEARCH METHODS 237
RESULTS 238
CONCLUSIONS 243
ACKNOWLEDGMENT 244
REFERENCES 244
CITATION 245

Index 247

CHAPTER 1

Unmanned Aerial Systems for Photogrammetry and Remote Sensing: A Review

I. Colomina , P. Molina

Centre Tecnològic de Telecomunicacions de Catalunya, Carl Friedrich Gauss 7, 08860 Castelldefels, Spain

ABSTRACT

We discuss the evolution and state-of-the-art of the use of Unmanned Aerial Systems (UAS) in the field of Photogrammetry and Remote Sensing (PaRS). UAS, Remotely-Piloted Aerial Systems, Unmanned Aerial Vehicles or simply, drones are a hot topic comprising a diverse array of aspects including technology, privacy rights, safety and regulations, and even war and peace. Modern photogrammetry and remote sensing identified the potential of UAS-sourced imagery more than thirty years ago. In the last five years, these two sister disciplines have developed technology and methods that challenge the current aeronautical regulatory framework and their own traditional acquisition and processing methods. Navety and ingenuity have combined off-the-shelf, low-cost equipment with sophisticated computer vision, robotics and geomatic engineering. The results are cm-level resolution and accuracy products that can be generated even with cameras costing a few-hundred euros. In this review article, following a brief historic background and regulatory status analysis, we review the recent unmanned aircraft, sensing, navigation, orientation and general data processing developments for UAS photogrammetry and remote sensing with emphasis on the nano-micro-mini UAS segment.

INTRODUCTION

Let them fly and they will create a new market. Of course, we are referring to Unmanned Aerial Systems (UAS) and UAS-based aerial remote sensing and mapping. According to recent market research (MarketsandMarkets, 2013) the global unmanned aerial systems market revenue is worth 5400.0 M€ as of 2013 and expected to grow up to 6350.0 M€ by 2018. *Let them fly and they will create a new remote sensing market in your country* is the message this paper would like to send to local policy makers, regulatory bodies and mapping authorities all around the world. Entrepreneurship together with robotic, computer vision and geomatic technologies have established a new paradigm (Colomina et al., 2008) of aerial remote sensing and mapping that, for some years now, has been serving the needs of large-scale low-altitude imaging and geospatial information users and developing an industry of its own (Cho et al., 2013, Mayr, 2013 and Petrie, 2013). The topic has become so important that the European national mapping agencies have organized working groups and begun to establish a common position (Cramer et al., 2013). Perhaps, we are standing right in front of the doors of the future, as suggested by the pioneering case of Trinidad and Tobago, where the government has already issued a tender for an UAS system "to provide colored map and imagery products to serve the identical purpose of conventional aerial survey and to satisfy the needs of a demanding spatial data market." GORTT (2013) Quite a forward looking decision. Yet, let us first explore what brought us here.

On names and acronyms

UAS are known under various different names and acronyms, such as "Unmanned Aerial Vehicle" (UAV), "aerial robot" or simply "drone," with "UAV" and "drone" being the most popular terms. The term UAS was adopted by the US Department of Defense (DOD) and the Civil Aviation Authority (CAA) of the UK. The International Civil Aviation Organization (ICAO) has introduced the concept of "Remotely-Piloted Aerial System" (RPAS), a particular class of UAS, in the ICAO Circular 328 (ICAO, 2011). This term is basically motivated by the fact that only RPAS will be able to integrate into the international civil aviation system. The aforementioned circular is a key reference in which the reader may find a comprehensive compilation of terms and definitions associated to UAS.

In this paper, we will refer to UAS for the system comprising an unmanned aircraft (UA), a ground control station (GCS) and a communications data link for the UA command and control (C2) from the GCS.

Pioneers

UAS were born (A.M. Low's "Aerial Target" of 1916; the Wright brothers Hewitt-Sperry Automatic airplane also in 1916) and raised (the Royal Navy used the Queen Bee drone for gunnery practice in 1933; USAF Firebees were used in North Vietnam and also by Israel against Egyptian targets in the Sinai during the 1973 October War) in the military context. Yet, the mapping potential of unmanned platforms was already understood by research groups in the late nineteen-seventies (Przybilla and Wester-Ebbinghaus, 1979 and Wester-Ebbinghaus, 1980). Navigation and mapping sensors were integrated onto radio-controlled platforms to acquire low-altitude, high-resolution imagery. The idea did not find many enthusiasts in the academic community, as shown by the limited number of publications and conferences. However, visionary technology and service companies that were well aware of their user needs, and open-minded civil aviation authorities that anticipated the social and business benefits of unmanned aircraft, soon started to develop, apply and regulate the technology (Petrie, 2013). Remarkable examples of this can be found in Australia, Japan and the UK. But how far did those initial developments make it through the thick jungle of differing user requirements, inexistent regulations and constant technological evolution?

General evolution

We live in the information century, and Internet is one of its main drivers. This constitutes grounds for the following minor exercise. In September 2013, more than six million entries were found in Google when searching for the words "Unmanned Aerial Vehicles," and almost twelve million when searching for its acronym, UAV. Using Google Trends, one sees that Internet usage of the word UAV in 2013 has diminished to almost half of the amount registered in 2005. Yet, this might be partially explained by the new, popular term "drone." As a matter of fact, there has been a clear rise in the use of this word since 2009, reaching around ten times its 2005 level in 2013 (the maximum peak was achieved in December 2011, when the US government asked Iran to return a lost drone). This analysis is a simple, non-quantitative, yet fairly illustrative approximation to measuring the impact of UAS in current times.

In the attempt to quantify the evolution of UAS development and its penetration into current professional markets, it may be useful to analyze the number of inventoried UAS as a direct indicator of how their importance has grown. Table 1 details the number of UAS systems referenced in the 2013 annual inventory of UVS International (van Blyenburgh, 2013). This table is an extension of the review work presented in Everaerts (2009). UVS International represents manufacturers of Unmanned Vehicle Systems (UVS), subsystems and critical components for UVS and associated equipment, as well as companies supplying services with or for UVS, research organizations and academia. The annual reports are reference materials on UAS inventories, and will be further mentioned in our contribution.

Table 1. Number of referenced UAS, developmental initiatives and purpose, period 2005–2013.

Item	2005	2006	2007	2008	2009	2010	2011	2012	2013
Referenced UAS	544	603	789	974	1190	1244	1424	1581	1708
Producers/ developers	207	252	312	369	422	500	511	478	540
Int'l teamed efforts	20	32	34	35	38	36	54	32	48
Producing countries	43	42	48	48	50	51	51	51	53
Devlopment purposes									
Civil/commer cial	55	47	61	115	150	171	175	217	247
Military	397	413	491	578	683	631	674	548	564
Dual purpose	44	77	117	242	260	283	318	353	392
Research UAS	35	31	46	54	66	66	69	73	78
Developmenta l UAS	219	217	269	293	329	301	310	187	172

Among the many interpretations that one may extract from the above table, an interesting trend is revealed: the number of developed UAS has multiplied by three from 2005 to present and, additionally, a relevant increase is observed in the civil/commercial type of platforms, especially in 2012 and 2013. Thus, it seems that PaRS UAS (clearly framed within that group) is cradled in a growing niche.

Literature evolution

Let us now focus on the scientific impact of UAS by screening the number of published papers at some of the most important PaRS conferences, for example the quadrennial International Society for Photogrammetry and Remote Sensing (ISPRS) congress. In 2004, the ISPRS congress in Istanbul hosted three UAS-related papers but did not feature any session specifically devoted to unmanned platforms. The trend changed in 2008, in Beijing, where 21 papers related to the use of UAS for PaRS and mapping purposes were presented in three different sessions. At the recent ISPRS congress in Melbourne in 2012, nine sessions related to UAS were held, featuring around 50 UAS-related papers. The international photogrammetric community has set up a dedicated biennial conference that began in 2011: the UAV-g (UAV-g 2011 in Zürich, Switzerland, UAV-g 2013 in Rostock, Germany and the upcoming UAV-g 2015 in Toronto, Canada). The increase in UAS-related publications at these conferences is clear, yet not exclusive. The IEEE Geoscience and Remote Sensing Society (IGARSS) has featured UAS-related papers at its annual symposiums since 2005. UAS-related papers have also been presented at the American Society for Photogrammetry and Remote Sensing (ASPRS) congresses, from 2005 in Baltimore up to present editions. Furthermore, the Multidisciplinary Digital Publishing Institute (MDPI) Open Access Journal of Remote Sensing published a special issue called "Unmanned Aerial Vehicles (UAVs) based Remote Sensing," closed in June 2012, with around 12 peer-reviewed papers. The IEEE Transactions on Geoscience and Remote Sensing journal also compiled seven papers on the use of UAS for Earth observation, published on 2009, and has been publishing UAS-related papers since 2007. The "Photogrammetrie, Fernerkundung und Geoinformation" (PFG) journal has featured five papers since 2007, with three in 2012.

A complete quantification of the UAS impact on current scientific disciplines should include a report on the number of papers in the conferences and journals of robotics and computer vision, like those sponsored by the IEEE Robotics and Automation Society (IEEE RAS) and the computer vision community respectively. These two disciplines have empowered UAS technologies and have come into the PaRS field with their own developments and ideas. The IEEE RAS organises and/or sponsors many UAS conferences. Among them, three ones deserve explicit mention: the IEEE International Conference on Robotics and Automation (ICRAS), the IEEE International Conference on Automation Science and Engineering (CASE) and the IEEE/Robotics Society of Japan (RSJ) International Conference on Intelligent Robots and Systems (IROS). As for IEEE journals, UAS technology can be found in the IEEE Transactions on Automation Science and Engineering (IEEE TASE), IEEE Transactions on Robotics (IEEE TR) and the IEEE Robotics and Automation Magazine (IEEE RAM). Far from being overlooked, its statistics are omitted here as the focus of the paper is PaRS.

Organization of the article

After providing an outline of how interest in UAS for PaRS has evolved, we introduce some history of the use of UAS in PaRS (Section 2), to later provide an overview of the current status of the main UAS technologies and regulations (Sections 3 and 4). In Section 5 we review the navigation, orientation and remote sensing payloads and in Section 6 we address UAS-sourced data post-processing. We conclude the article with a review of UAS geomatic applications and markets.

EARLY DEVELOPMENTS

Balloons are probably the oldest platforms for aerial observation. As a matter of fact, aerial photographs of Paris were already being captured in 1858 by Tournachon aboard a hot-air balloon. Later on, and thanks to the simplification of camera technology, other means such as kites (used by the English meteorologist E.D Archibald in 1882) and rockets (as used by the Swedish inventor Alfred Nobel in 1897) were used for aerial photography. Perhaps one of the most exciting early experiments was the use of small cameras mounted on the breasts of pigeons from the Bavarian Pigeon Corps, as proposed by J. Neubronner in 1903. Thus, Tournachon's adventures aside, one may conclude that the oldest form of aerial remote sensing was actually based on remotely-piloted vehicles. Yet, the reader would be right to harbor doubts about defining a pigeon as an "unmanned platform" or "remotely-piloted aircraft".

Manned airborne aerial photographs came later (in 1909 W. Wright shot a motion picture aboard his home-made airplane), and rapidly became well-established tools in the military field, mainly for war purposes given the context of Europe at that time. It is out of the scope of this paper to comment on how photogrammetry was born and evolved, in the aerial or satellital platform context as well as the imaging sensor context.

Indeed, the evolution of integrated circuitry and radio-controlled systems in the late twentieth century was key in the advent of modern UAS for PaRS. In 1979, Przybilla and Wester-Ebbinghaus performed a test with a radio-controlled, fixed-wing UAS with a length of 3 m and equipped with an optical camera (Przybilla and Wester-Ebbinghaus, 1979). A second test was carried out in 1980 by the same team but using model helicopters carrying a medium-format Rolleiflex camera (Wester-Ebbinghaus, 1980). The prototype was used to aerially document an old steel construction, and it was first use of rotary-wing platforms for PaRS. From then until present times, the rotary- or fixed-wing, single- or multi-rotor, remotely- and/or auto-piloted platforms have been established in most of the UAS implementations for PaRS (those early experiments truly paved the way for future developments). The reader interested in the early stages of UAS-based PaRS is referred to the comprehensive literature review in Eisenbeiss (2009), and also to a comprehensive review of ancient and modern aviation history and unmanned flights in Dalamagkidis et al. (2009).

UNMANNED AERIAL SYSTEMS AND UNMANNED AERIAL SYSTEMS FOR PARS

A UAS is a system of systems—that is, a set of complementary technologies brought together to fulfil a specific task—and, as such, there currently exists a wealth of different systems: it may be said that there is one for every combination of technologies. At the highest level of the UAS technology breakdown, three main UAS components are commonly identified, namely the *unmanned aerial vehicle*, the *ground control station* and the *communication data link*. Further down, other UAS components are considered critical, such as autopilots, navigation sensors, imaging sensors, mechanical servos, and wireless systems. In this paper, some of these UAS technologies with relevance within the PaRS field are presented and surveyed.

The categorization of the existing unmanned aircraft has been a constant and necessary exercise among the UAS community throughout its history "to bring order into chaos," quoting from the EuroSDR report of the project "NEWPLATFORMS" (Everaerts, 2009). As a result of this need, there is a myriad of classifications of UAS, according to the various characteristics of the aerial platform (size and weight, endurance, aerodynamics, etc.) or the system operation (mission range or flying altitude, nature of its application, etc.). We review some of current work on UAS classification hereafter.

The work presented in Eisenbeiss (2009) with respect to UAS categorization is noteworthy; it considers powered and non-powered, heavier- or lighter-than-air platforms and performs an assessment on range, endurance, weather and wind dependency, and maneuverability. It additionally defines a proprietary classification based on price and payload, including on-board navigation sensor grade and, thus, geo-referencing or real-time capabilities, and application accuracy requirements.

In van Blyenburgh (2013), a vast inventory of world-wide UAS, including categorizations based on size, weight, operating range and certification potential is provided. The classification based on size and operation range, depicted in Fig. 6, p. 169, with the corresponding number of inventorised platforms is of special interest. This table reveals three fairly well-distinguished UAS ecosystems: *nano-micro-mini* UAS, *close-short-medium-range* UAS, and the rest of UAS, which we will discuss hereafter.

From the last to the first, the Medium Range Endurance to Exo-Stratosferic UAS ecosystem groups the largest UAS with highest operating altitudes, which often consist of fine and complex avionics and are only allowed to fly under certain special regulations by certain specific teams, generally military units. Together, they total up to 179 referenced UAS. Secondly, *close-short-medium-range* UAS are characterized by an Maximum Take-Off Weight (MTOW) between 150 and 1250 kg and an operating range between 10 and 70 km. This group includes fixed- and rotary-wing developments from well-established companies (Raytheon, Northrop Grumman, Saab) generally equipped with remote sensing technologies, and total up to 546 developments. Finally, the *nano-micro-mini* UAS class is defined by low weights and payload sizes, low flying altitudes and quick operational deployments, and concentrates up to 728 developments. More specifically, *mini* UAS, which are defined by an operative range of less than 10 km, allowed to fly lower than national ceilings of segregated airspaces, feature less than two hours of endurance and less than thirty kilograms of MTOW (although the MTOW specification may vary from country to country), is the largest group, featuring 490 referenced systems. This UAS ecosystem seems to fit the PaRS community needs, in view of the existing PaRS UAS to be reviewed throughout this article.

At this point, a trade-off is evident: on the one hand, there is not a unique and universal classification of UAS but, on the other, it is necessary to distinguish among the myriad of platforms already existing in the world. In order to set a basis for our paper, we will use the following weight-based categorization: micro (less than 5 kg); mini (less than 30 kg); and tactical (less than 150 kg). With these categories, we aim at simplification and ease of reading, as well as covering the UAS portion of interest for the PaRS community. Additionally, based on our own experience, these names and definitions are fairly common among UAS professionals.

We will now review the crucial high-level components of a UAS, focusing on PaRS.

Unmanned aircraft

Conditioned by [the lack of] regulations and, to a certain degree, by the inherent complexity of large systems, a UAS PaRS mission has usually consisted of operating an aerial platform, most probably a fixed-wing or rotary-wing craft of less than 30 kg MTOW, within a range not greater than 10 km and flying below 300 m, carrying a small or medium-format optical camera (probably on the visible spectrum), and either remotely piloted by a human or automatically piloted by an autopilot based on two main navigation technologies, Global Navigation Satellite Systems (GNSS) (such as for example GPS) and Inertial Navigation Systems (INS), in a GNSS-only mode or INS/GNSS coupling. In our opinion, this is a representative picture of current unmanned aircraft for PaRS.

We believe that, in general, the geomatic community has finally taken positions on the*nano-micro-mini* UAS ecosystem. Yet, the history of UAS PaRS is marked by a heterogenous array of developments that sometimes departing from the previous ecosystem. Everaerts et al. (2004) presented a stratospheric UAS to offer high resolution aerial imagery in near real time for large-scale mapping and crisis monitoring, and presented recent updates in Everaerts and Lewyckyj (2011) with respect to the integration with local civil aviation authorities. More recently, Miraliakbari et al. (2012)reported on the use of gyrocopters as potential mapping platforms, focusing on the analysis of its vibrations, and Thamm (2011) presented a parachute-based remote-sensing system. In Kemper (2012), several non-conventional unmanned and manned platforms (balloons, blimps, trikes and paratrikes) are presented carrying remote-sensing payloads.

Everaerts (2009) features a comparative analysis of airborne platforms, satellite platforms, low-altitude and high-altitude UA, in relation to their main qualities for remote sensing missions, such as coverage, update rate, flexibility, quality, spatial resolution, positional accuracy, spectral resolution and accuracy, target applications and system economic cost. An update on the use of UAS for geomatics, covering market and legislation analysis, is provided in Haarbrink (2011).

In order to provide the reader with an idea of the current unmanned aircraft used for PaRS, and without attempting to provide a comprehensive list, a few systems are described in Appendix A. Table 10 compiles the main characteristics of micro and mini fixed-wing, rotary-wing and multi-rotor UAS for PaRS, respectively.

Ground control station

As per common understanding, Ground Control Stations (GCSs) are stationary or transportable hardware/software devices to monitor and command the unmanned aircraft. Although the word *ground* is inherent to the concept, a UA may actually be operated from the ground, sea or air. GCS are probably as important as the unmanned aircraft themselves, as they enable the interface with the "human intelligence"—any change in the route of the UAS, any eventual error on the aerial platform and/or any outcome of the payload sensors shall be sent to and seen within the GCS. As fundamental pieces in UAS, GCS have evolved over the past decades pushed by the parallel improvements in computer science and telecommunications. van Blyenburgh (2013) provides a compilation of referenced GCS, most of which are military.

Requirements in UA-GCS communication, commanding devices, the number of monitors and the crew members needed to command a UA are crucial variables to shape a particular GCS. From the Predators or GlobalHawks GCS, enabling simultaneous control of several platforms and command by up to six crew members, to small portable PC-based GCS such as that in the UX5, from Trimble, down to the software-only GCS, found in SwingletCAM, from SenseFly.

Even though, generally speaking, a commercial UAS such as any of those listed in Table 10 is a non-separable UA-GCS ensemble, there are some generic developments usable as stand-alone solutions, such as the Portable Ground Control Station, from UAV Factory.

Communication

UAS communication is critical in terms of mission requirements (that is, to command and control the aircraft and eventually screen the payload outcome) as well as safety, especially when it comes to UAS integration with Air Traffic Control (ATC) in non-segregated airspace.

The debate on UAS-related communication issues was present at the World Radiocommunication Conference (WRC-12), held in Geneva in February, 2012. As published on the International Telecommunication Union (ITU) website, the aim was to consider spectrum requirements and possible regulatory actions, including the identification of a globally harmonized spectrum, in order to support the safe operation of unmanned aircraft systems in the non-segregated airspace used by civil aviation (agenda item 1.3). Although unmanned aircraft systems have traditionally been used in segregated airspace where separation from other air traffic can be assured, administrations expect broad deployment of unmanned aircraft systems in non-segregated airspace alongside manned aircraft. The outcomes of the WRC-12, particularly focusing on agenda item 1.3, consisted of a new allocation to Aeronautical Mobile Satellite (Route) Service (AMS(R)S) in support of UAS, agreed to in the 5030–5091 MHz band, which is the core band for Microwave Landing Systems (MLS). The ITU Radiocommunication Sector was instructed to conduct studies to develop technical, regulatory and operational recommendations for the WRC in 2015. As agreed future actions, the ITU-R and WRC-15 will explore whether the Fixed Satellite Service (FSS) can be used to provide UAS command-and-control communications, consistent with the aeronautical safety of flight requirements, and highlighting the need for participation of aviation safety regulators and other aviation interests.

Indeed, when no integration with ATC is required (e.g. operation in segregated airspace[1]), communication links are only subject to standard frequency spectrum legislation. Thus, the responsibility for safety falls on the side of the particular radio-communication technology implemented on the UAS.

Many communication technologies are used in today's UAS, the most predominant of which in the Mini UAS category is Wi-Fi (usually around 2.4 GHz), as is used by some of the UAs in Table 10. Yet, other technologies have been considered, such as high-frequency satellite communication as in military systems (Predator, Global Hawk) or Worldwide Interoperability for Microwave Access (WiMAX), an interoperable implementation of the IEEE 802.16 wireless standard (Dusza and Wietfeld, 2010). A comprehensive overview of aspects related to command, control and communication technologies for small UAVs is provided in Barnard (2007). In addition, this presentation discusses aspects of the integration of UAS into non-segregated airspace and provides a comparison table of the frequencies used by some existing UAS, such as ScanEagle, FireScout or RMAX.

Mission planning

In the formal definition of UAS—aircraft, control station and data link—there is no explicit mention of mission preparation and execution. However, an essential component of UAS for geodata acquisition is the mission planning and management subsystem. Experience shows that a careful design of the aircraft trajectory (waypoints, strips, speed, attitude, etc.) and a flexible real-time mission management capacity (sensor configuration, triggering events, flying directions, etc.) are instrumental in achieving productive and safe acquisition missions (Mayr, 2011b).

Although the mission planning and real-time management component is usually an integrated part of commercial UAS (e.g., Micropilots Horizon or QGroundControl used by ArduPilot), there are open issues that still lead users to repeat acquisition campaigns simply because of flawed mission design or execution. Apparently, the "easy" operation of micro- and mini-UAS should mitigate this inconvenience However the aircraft of micro- and mini-UAS are tiny vehicles, sensitive to wind and wind bursts and of limited autonomy. A typical feature, for instance, of mission plans for UAS photogrammetry is the large forward (80%) and cross (60–80%) overlap to compensate for aircraft instability. A rare feature, for instance, of mission management is to accommodate flight plans to the actual wind conditions of the mission area at the time of mission execution. Mission planning and real-time mission management subsystems are the key to a competitive exploitation of UAS for photogrammetry and remote sensing. An illustrative example of a real-time mission management subsystem is described inStødle et al. (2013) where extreme operational conditions and long distances do not allow for mission failures. The system features real-time and off-line visualization of UAS-sourced images and therefore a quality-control tool on data completeness, etc. so the remote crew can take control of the unmanned aircraft and correct any possible deviations from the mission plan.

REGULATORY BODIES AND REGULATIONS

The implementation of a harmonized regulatory panorama has been demanded by UAS stakeholders to drop the barriers for UAS certification and commercialization all over the world. Yet, this issue entails non-trivial technicalities and involves a large group of contributing agents. Based on the comprehensive list provided in Everaerts (2009), we present a summary of the main actors in UAS regulations and latest updates to date of publication.

The ICAO brings together states and key industry organizations to develop policies and standards on all aspects of civil aviation activity. In 2007, the Unmanned Aerial Systems Study Group (UASSG) was created to be the focal point and coordinator of all ICAO UAS-related work. As already mentioned in the first chapter of this article, the Circular 328 on UAS was released in 2011 by ICAO, being the first official ICAO document on the subject. Some relevant statements from this document can be highlighted, such as the fact that any UAS-related incident shall be investigated by the competent aviation bodies, and also the possibility to certify the aircraft and the ground station separately, which may be of high relevance to industrial stakeholders.

The European Aviation Safety Agency (EASA) is a European Union (EU) agency, gathering the civil aviation authorities of the member States, which developed a policy statement on "Airworthiness Certification of Unmanned Aircraft Systems (UAS)" in August, 2009. As published on its website, "this policy establishes general principles for type-certification (including environmental protection) of an Unmanned Aircraft System (UAS)."

The European Organisation for the Safety of Air Navigation, EUROCONTROL, is devoted to the creation of a uniform Air Traffic Management (ATM) system gathering civil and military users. In 2007, EUROCONTROL released the "Specifications for the Use of Military Remotely-Piloted Aircraft (RPA) as Operational Air Traffic outside Segregated Airspace" and has been updating it until the present time (the last update of this document is from February 2012).

On the defence and military side, other European actors contributing to regulations are the European Defence Agency (EDA), which aims to support the member states on improving European defence capabilities, and the North Atlantic Treaty Organization (NATO), responsible for the military regulations of most of the European countries. The EDA has been focusing on UAS air traffic insertion, frequency management and future UAS, and has funded several initiatives to its respect such as the AIR4ALL project (2008). On the other side, NATO has participated in a regulatory process through its Flight In Non-segregated Airspace (FINAS) group, although it clearly states they are not a regulatory body on this issue. The UAS-related work of NATO spans many aspects of UAS management, namely operator training, airworthiness, risk assessment, system architectures, etc., and has produced a wide collection of standards of best practices.

Additionally, there are a few noteworthy initiatives in the non-profit, user-driven category. First, the Association for Unmanned Vehicle Systems International (AUVSI), which is "the world's largest non-profit organization devoted exclusively to advancing the unmanned systems and robotics community" as stated on their website. Second, the European Organisation of Civil Aviation Equipment (EuroCAE), which deals exclusively with aviation standardisation (airborne and ground systems and equipment) and related documents as required for use in the regulation of aviation equipment and systems. The EuroCAE WG-73 deals with the analysis of key issues related to UAS operations in the context of European ATM and UAS terminology and definitions.

Some research companies' initiatives have helped clear a path for advancement in UAS regulations. The first noteworthy example is Everaerts and Lewyckyj (2011), in which the interaction with real air traffic control is studied in depth and carried out. Schulz (2011)also presents an effort to understand regulations for UAS, and the development of a certifiable UAS system admitting flexible scientific payloads.

Recent events and decisions are proof of the active work carried out by competent authorities on UAS regulation. For example, the US Federal Aviation Administration (FAA) was required to develop a "comprehensive plan for integrating UAS into the national airspace system no later than September 30, 2015" as stated in the 2012 FAA Reauthorization Act. In response to the mandate, the FAA recently published the "Integration of Civil Unmanned Aircraft Systems (UAS) in the National Airspace System (NAS) Roadmap" (Federal Aviation Administration, 2013). In the document, an initial "accommodation" phase and a final "integration" phase are described that, altogether, will last much longer than 2015. FAA conditions for the UAS being integrated into airspace is that they cannot reduce capacity, decrease safety, increase the risk (on air and on ground) and impact operations any more than other new technologies. Quoting (Federal Aviation Administration, 2013) "Integration efforts will focus on sequentially developing and implementing the UAS system requirements established by the FAA as a result of Research and Development and test range outputs." For the latter purpose the FAA has already selected six test sites (Davis, 2013). For PaRS applications it is interesting to note that, among other general rules to abide by, the FAA Roadmap announces that "all UAS will require design and airworthiness certification to fly civil operations in the NAS" with the exception of "some special cases, such as small UAS (sUAS) with very limited operational range."

Last but not least, in the EU regional context, the European RPAS Steering group (ERSG) released (June 2013) the "Roadmap for the integration of civil Remotely-Piloted Aircraft Systems into the European Aviation System" a comprehensive report that proposes a plan spanning the period from 2013 to 2028 (ERSG, 2013). It includes detailed proposals and a schedule for a regulatory approach, a strategic research plan and a study on the societal impact of UAS. The ERSG is aware that most companies active in this business sector are small and medium enterprises (SMEs) which would be unable to cope with a disproportionate regulatory burden; and that, in addition, disproportionate regulation would considerably reduce the potential of UAS technology to develop a new market. In this respect, and in order to create a seamless EU UAS market, the ERSG report also proposes the transfer of national competences for UA under 150 kg to the EU by 2016. In the meantime (and this is of special interest for UAS PaRS applications since the majority of UAS commercial applications are undertaken with light UA), the development of regulations remains in the hands of the national CAAs. Up to now, fifteen EU countries have developed partial regulations for civil UAS operations and another five are in the process of doing so.

In Australia, as early as 2002, the Civil Aviation Safety Authority (CASA), passed Civil Aviation Safety Regulation 101 (CASR 101) (CASA, 2002), the first operational regulation for unmanned aircraft in the world. Now, thanks to CASR 101, anyone interested in becoming a UAS operator—those flying UAS for business in contrast to recreational and sport users—benefit from a regulatory framework that allows them to apply for a UAS operator's certificate. The certification covers a wide range of aspects including UAS controllers, UAS pilots, and medical, maintenance, and liability areas among others. At the time this paper was written, there were 40 certified UAS operators and of them 29 were certified for aerial survey operations (CASA website).

In the UK, the Civil Aviation Authority (CAA) has regulated UAS flights through the "Air Navigation: The Order and the Regulations" (CAP 393) and "Unmanned Aircraft System Operations in UK Airspace" (CAP 722) documents respectively (CAA, 2012a and CAA, 2012b). A key asset to UAS operations is the Basic National UAS Certificate (BNUC), the CAA's accepted qualification for lightweight UAS pilots and crews. BNUCs are managed and awarded by the European Unmanned Systems Centre (EuroUSC) a UK organization that has been authorized by the CAA to asses the airworthiness of lightweight UAS of 150 kg and under. Of special interest for UAS-based photogrammetry and remote sensing is the BNUC-S, for the operation of UAS with a Maximum Take Off Mass (MTOM) of less than 20 kg that are used for "aerial work" under Visual Line of Sight (VLOS) or Extended VLOS (EVLOS) operational conditions.

In parallel, on a local and regional, partial, although encouraging steps, are being taken. In October 2012, the French Civil Aviation Authority, the DGAC, granted the first authorization ever issued to a UAS company in France to Delair-Tech, for a civil UAS to fly over 100 km in French airspace. This is of major relevance for power and pipe line surveys and, in general, any Beyond Visual Line of Sight (BVLOS) operation. In December 2012, COWI, the Danish photogrammetric and mapping company was the first private company in Denmark to be authorized to use UA for aerial surveys of disasters and other tasks that can benefit from aerial images in both urban and rural areas without further approval. At the time this article was written, there were a total of seven organizations authorized by the Danish Transport Authority (*Trafikstyrelsen*) to fly UA. If the survey area falls within a radius of five miles of an airport, the airport must be notified of the flight. In April 4, 2013, the company Aermatica conducted the first authorized UAS flight in non-segregated Italian airspace with its unmanned Anteos system. The authorization was granted by the Italian Civil Aviation Authority (*Ente Nazionale per l'Aviazione Civile*, ENAC) to perform an aerial survey of the *Basilica di S.M. di Collemaggio* in L'Aquila and other churches damaged during the L'Aquila earthquake.

In May 2013, the Brazilian National Civil Aviation Agency (*Agência Nacional de Aviação Civil*, ANAC) has recently authorized private operators to fly UA. Up to now, the Federal Police was the only non-military organization allowed to do so. X-Mobots (São Carlos, SP), the first private company to benefit from the decision, obtained its Certificate of Experimental Flight Authorization (*Certificado de Autorização de Voo Experimental*, CAVE) from the ANAC.

For further reading on UAS regulations and regulation roadmaps, in addition to the ERSG (ERSG, 2013) and FAA reports (Federal Aviation Administration, 2013), a comprehensive description on the most recent work is presented in van Blyenburgh (2013), including the ICAO circular 328, reports on the EuroCAE WG 73 on UAS and the experiences of different expert groups from several countries around the world.

NAVIGATION, ORIENTATION AND SENSING PAYLOADS

For PaRS applications, two critical components of a UAS are the navigation-and-orientation payload and the remote-sensing payload.

In a typical UA, the "autopilot" loop repeatedly reads the aircraft's position, velocity and attitude (tPVA, with t standing for time) from the Navigation System (NS) and uses the tPVA parameters to feed the Flight Control System (FCS) to guide the aircraft (Elkaim et al., 2014). Specially in PaRS UAS, an Orientation System (OS) is also in place to estimate the same tPVA parameters but not necessarily in real-time e.g. to perform a posteriori sensor orientation for mapping. In an "ideally"-designed UA for PaRS, the NS sensors—inertial measurement unit (IMU), GNSS receiver, baroaltimeter, compass and possibly others—would be shared by the OS or the redundancy of the NS and OS sensors would be exploited. In practice, most times, the NS and OS are separated or the NS of the autopilot provides the tPVA solution to the OS. In the latter case the OS only has a trivial input/output function and its orientation parameters are not used as aerial control but as initial approximations for the automatic generation of tie points and their photogrammetric measurements. This is so because the NS is required to provide a real-time high-frequency (up to 1 kHz) tPVA solution of low to moderate accuracy as opposed to the OS that is required to provide a high accuracy tPVA solution although it is post-processed and at lower frequency. This chapter compiles information on available, commercial NS and OS suitable for UAS PaRS. Information supplied by the manufacturers' brochures and websites is presented in tables for both categories.

Autopilots and navigation systems

The components listed in Table 2 are just a few examples from the wide spectrum of UAS autopilots. Note that autopilot weights do not include GPS antennas (which may be the limiting weight factor for micro-UAS). Also note that the presented products may admit additional sensors and/or demonstrate different capabilities. AirWare, formerly Unmanned Innovation Inc. (USA), provides a range of autopilots from closed solutions to customized devices, featuring integrated INS/GPS navigation and flight control systems. MicroPilot (Canada) has been producing a wide range of products since 1995, which have been used by companies or institutions such as SRS Technologies (USA), BlueBird Aerosystems (Israel), INTA (Spain) and several branches of NASA. Prices range from 2000 to 8000 US dollars, multiplying the on-board computational capabilities by fifty times (in higher-end models). Piccolo, by CloudCap Technologies (USA), provides a solution including the core autopilot, flight sensors, navigation, wireless communication and payload interfaces, and has been used for PaRS (Schinstock et al., 2009 and Cosentino, 2008). UAV Navigation describe themselves as providers of "high performance navigation and autopilot solutions used by many Tier1 aerospace companies in a wide range of UAS such as high-performance tactical unmanned planes, targets, mini UAS, helicopter and quadcopter aerial vehicles," and has also been used by PaRS groups (Zarco-Tejada and Berni, 2012). For additional information, a survey of UAS autopilots can be found in Chao et al. (2010) and a comprehensive compilation of the main brands of UAS autopilots is provided in UAVMarketSpace (2012). Finally, van Blyenburgh (2013) compiles a list of 201 inventorised autopilots.

Table 2. Autopilot examples.

Product/company	Sensors	Weight (g)	Size (cm)	Platform	Price (kE)
osFlexPilot/Airware	INS/GPS/AS/MM	200	11 × 5.5 × 2.8	Fixed-wing, VTOL	⩽5.5
osFlexQuad/Airware	INS/GPS/AS/MM	32	6.73 × 6.73 × 2	Multi-rotor	⩽4
MP2128/MicroPilot	INS/GPS/BA/AS	24	10 × 4 × 1.5	Fixed-wing, VTOL	⩽4.4
MP2028/MicroPilot	INS/GPS/BA/AS	28	10 × 4 × 1.5	Fixed-wing, VTOL	⩽2.5
PiccoloNano/CloudCap Technologies	GPS/AS/BA	65	4.6 × 7.6 × 2	Small platforms	–
Piccolo II/CloudCap Technologies	INS/GPS/BA/AS	226	14.2 × 4.6 × 6.2	Fixed-wing, VTOL	–
VECTOR/UAV Navigation	INS/GPS/MM/BA/AS	180	–	Mini and Tactical	–
ArduPilot Mega 2.5/3DRobotics	INS/MM/BA	17	6.7 × 4 × 1	Fixed-wing, VTOL	0.12

AS: airspeed sensor; BA: baroaltimeter; MM: magnetometer; weigths do not include GPS antennas.

ArduPilot Mega 2.5 does not include GPS receiver, power module and telemetry module.

Table 3. Commercial Hybrid Measurement Units (HMU) and Hybrid Navigation Systems (HNS) for unmanned aircraft orientation.

Make	Model	Weight (kg)	GNSS Φ or ρ	σ_P (m)	Weight IMU (kg)	σ_a ($\mu G/\sqrt{Hz}$)	μ_a (μG)	σ_ω (deg/s/$\sqrt{Hz}$)	μ_ω (deg/h)	$\sigma_{\theta,\gamma}$ (deg)	σ_ψ (deg)
iMAR	VRU-FQ	1.750	Φ	0.05–0.30	1.750	50	100	0.0016	11	NA	NA
Applanix	AP LN-200 E	0.250	Φ	0.05–0.30	0.750	50	50	0.0008	0.5	0.008	0.025
Novatel	SPAN-IGM	0.515									
Adv. Nav.	Spatial-Dual	0.304	ρ	1.20–2.00	Included	246	35	0.009	4	0.15	0.10
Novatel	MIC ADIS-16488	0.124	Φ	0.01–0.02	0.048	67	100	0.005	6	0.04	0.22
Independent test										0.09	0.16
iMAR	iμVRU-01	0.050	ρ	1.20–2.00	Included	310	2000	0.005	10	0.20	<1
Adv. Nav.	Spatial	0.025	Φ	0.01–0.02	Included	400	60	0.005	18	0.40	0.80
Independent test			ρ	0.22–0.77						0.07	0.30
SBG	IG500-E (box)	0.049	Φ, ρ	0.05–0.30	Included	250	60	0.05	20	0.50	0.50
SBG	IG500-E (OEM)	0.010	Φ, ρ	0.05–0.30	Included	250	60	0.05	20	0.50	0.50

Φ: phase measurements; ρ: code measurements; σ_P: positional precision; σ_P: positional precision; σ_a: linear accelerations' noise (PSD level); μ_a: linear accelerations' in-run bias stability (PSD level); σ_ω: angular rates' noise (PSD level); μ_ω: angular rates' in-run bias stability

(PSD level); $\sigma_{\vartheta,\gamma}$: roll and pitch precision (whole spectrum); σ_ψ: heading precision (whole spectrum); NA: not available.

Several open-source frameworks exist to tackle auto-piloting in UAS. Meier et al., 2011 and Mészáros, 2011 present reviews of available open-source hardware and software for both UAS and UGVs, in addition to actual results based on in–house developments. Some noteworthy developments are the Paparazzi project (Brissset and Drouin, 2004), defined as "a free and open-source hardware and software project intended to create a powerful and versatile autopilot system for fixed-wing and multi-copter platforms," and already used in aerial imaging missions (Jensen et al., 2008 and Jensen et al., 2009). Other initiatives are the OpenPilot project (http://www.openpilot.org/about/), a community-driven open-source project combining software and hardware-plus-sensor solutions for fixed- and rotary-wing platforms, or ArduPilot (http://www.diydrones.com/notes/ArduPilot), created within the DIYDrones community, developing a family of open source autopilots based on the Arduino platform. The ArduPilot product family, manufactured by 3DRobotics, is quite popular among micro UAS users and also radio-control model aircraft developers, as it provides small and affordable products for guidance and control.

In addition, Unmanned Ground Vehicles (UGV) have both benefited and contributed to the advent of autopilot technology. Early developments such as CLARAty, by the NASA's Jet Propulsion Laboratory, preceded the recent Robot Operating System (ROS) (www.ros.org), which has been also adopted by aerial vehicles (Singh et al., 2012)

Orientation systems

Today's level of miniaturization of computer boards, GNSS receivers and antennas, IMUs and, in general, sensors allows for the integration of hybrid measurement units (HMU) for light unmanned aircraft whose measurements can be processed—in real-time, in a hybrid navigation system (HNS) or in post-processing, in a hybrid orientation system (HOS)—to deliver position orientation parameters at the cm-level (Rehak et al., 2013). These results are dominated by the quality of the GNSS receiver and, to an even larger extent, the quality of the GNSS receiver's antenna (van Diggelen, 2010). The accuracy of the attitude part of orientation is highly dependent on the IMU quality and flight dynamics, and it varies within the interval $[0.015, 0.2]$ deg for $\sigma_{\vartheta,\gamma}$ (roll and pitch) and within $[0.03, 0.5]$ deg for σ_ψ (heading). Therefore, the figures given in the next paragraph are simply indicative as they include general specifications and specific test results of different trajectories. To summarize the current capabilities of HMU and HOS, considering the critical role of the payload weight, we review a representative, though noncomprehensive, set of commercial systems with weights from around 1 kg down to 0.01 kg.

A 0.9–1.5 kg HMU (0.25 kg for its GNSS multiple-frequency antenna, 0.25 kg for its control unit including a GNSS multiple-frequency phase receiver and 0.4–1.0 kg for its IMU) guarantees a geodetic-grade tPVA post-processed solution (P: $\sigma_{e,n,u} < 0.05\ \mathrm{m}$, V: $\sigma_{ve,vn,vu} < 0.005\ \mathrm{m/s}$, A: $\sigma_{\theta,\gamma} < 0.015\ \mathrm{deg}$, $\sigma_{\psi} < 0.030\ \mathrm{deg}$). This category of HMUs and HNSs is represented by the Appanix (Trimble) AP family (AP20 and AP40) and Novatel's MEMS Interface Card (MIC) with tactical-grade or higher grade IMUs like the Northrop-Grumman LN200 or the KVH 1750. iMAR's iVRU-FQ belongs to this category although its weight (1.8 kg) is significantly higher. The performance of this technology is well-known and the figures above are repeatable.

For frame photographic cameras, the above precision performance is compatible with direct sensor orientation (DiSO)—a.k.a. direct georeferencing—for altitudes up to 50 m above ground and ground sampling distances (GSD) of 2 cm or larger. The heading precision requirement is dependent on the camera "size" (number of pixels) and shape (number of rows and columns) and for a consumer-grade camera of about 14 Mpx, for the mentioned height and GSD, a heading error precision of $\sigma_{\psi} < 0.06\ \mathrm{deg}$ would be sufficient. These are precision figures, which do not take into account the remaining INS/GNSS inaccuracies. For mapping applications, they assume that the camera is calibrated down to the 0.5 px (1-σ level) precision and accuracy. Therefore, although the performance is promising for DiSO, it is better suited to the Fast AT (Blázquez and Colomina, 2012a) and integrated sensor orientation (ISO) (Blázquez and Colomina, 2012b) procedures. Note that for DiSO, it is not just precision that counts but also accuracy; i.e., absence of systematic errors.

Advanced Navigation's Spatial-Dual HMU and HNS (0.304 kg without its two GNSS antennas, despite being in a shockproof case) implements an interesting concept: two GNSS geodetic-grade receivers featuring a triple-frequency capability, enabled for GPS, GLONASS, Galileo and BeiDou signals.

The double antenna design guarantees 0.1 deg heading accuracy regardless of the vehicle dynamics for 1 m or longer antenna-to-antenna distances. The HMU includes baroaltimeters, magnetometers, a MEMS IMU and optional odometers and pressure sensors for terrestrial and underwater navigation respectively. Based on a high-frequency IMU output rate, the Spatial-Dual can generate a tPVA solution up to 1 kHz making it ideal for both navigation and orientation systems in high-dynamic motion. According to the manufacturer, the Spatial-Dual can deliver positions at the cm level as provided by GNSS phase measurements, velocities better than 1 cm/s and attitudes at the 0.15 deg, 0.1 deg levels for $\sigma_{\vartheta,\gamma}$ and σ_{ψ} respectively.

These results allow for block adjustment with a reduced set of ground control points (due to the positional precision of aerial control). For very wide angle objectives and/or low resolution images—instantaneous field of view (IFOV) of 0.2 deg,—they are consistent with DiSO and sufficient for Fast AT. They also allow for fast automatic tie point identification and measurement.

MEMS-based, lighter HMUs at the 0.1 kg level, are represented by Novatel's combination of the MIC and the ADIS-16488 IMU from Analog Devices (0.08 kg without GPS antenna and protective box), iMAR's iIMU-01 and iVRU-01 units (0.05 kg without GPS antenna), Advanced Navigation's Spatial (0.025 kg without GPS antenna) and SBG's IG500-E (0.049 kg without GPS antenna, 0.01 kg without protective box for OEM integration). In Novatel (2013) the combination of Novatel's MIC and the ADIS-16488 IMU, is related to performances $\sigma_{e,n} \approx 0.01$ m, $\sigma_u \approx 0.02$ m, $\sigma_{\gamma,\vartheta} \approx 0.04$ deg, and $\sigma_{\psi} \approx 0.22$ deg that are consistent with an independent evaluation of the same IMU (Goodall et al., 2012) that delivered $\sigma_{\gamma,\vartheta} \approx 0.09$ deg, and $\sigma_{\psi} \approx 0.16$ deg. Advanced Navigation (2012) reports accuracies of $\sigma_{e,n} \approx 0.2$ m, $\sigma_u \approx 0.8$ m, $\sigma_{\gamma} \approx 0.05$ deg, $\sigma_{\vartheta} \approx 0.08$ deg and $\sigma_{\psi} \approx 0.3$ deg for its Spatial HNS in a terrestrial kinematic test (with odometer). However, the company is rather conservative in its navigation performance specifications ($\sigma_{\gamma,\vartheta} \approx 0.4$ deg, $\sigma_{\psi} \approx 0.8$ deg).

As with the Advanced Navigation's Spatial-Dual, cm-level positional precision for aerial control allows for reduced to minimal ground control configurations in integrated sensor orientation (ISO, block adjustment with position or position/attitude aerial control). This given, IMU quality and trajectory dynamics make the difference between just ISO or Fast AT and DiSO. dm-level positional accuracy can be exploited depending on GSDs and project requirements. Even in the poor accuracy case (m-level positioning, deg-level attitude), these orientation systems facilitate digital aerial triangulation as image connections through tie points can be rapidly derived from the known geometry between images.

Whether or not a low-cost OS whose results cannot be exploited as aerial control measurements is of interest is an open question. From a technical point of view it brings us back to the pre-GPS days of classical aerial triangulation. From a business point of view, achievement of sufficient productivity very much depends on local infrastructure (ground control point databases) and logistics (cost of ground surveys).

We conclude this section by quoting (Gakstatter, 2010): *as soon as 2014 and at latest by 2020, centimeter level accuracy will be in the hands of anyone with a few hundred dollars to spend*, which is consistent with the discussions and results in van Diggelen et al., 2011 and Rehak et al., 2013. Therefore, sooner than later, cm-level accuracies for exterior orientation parameters of images acquired with micro-unmanned aircraft and heavier aircraft will be an inexpensive, achievable goal. Geodetic-grade attitude precisions ($\sigma_{\theta,\varphi} < 0.015$ deg, $\sigma_{\psi} < 0.030$ deg) compatible with DiSO can be achieved with the traditional, "heavy" tactical-grade IMUs. Results of airborne missions combining geodetic-grade multi-constellation GNSS receivers with MEMS tactical-grade, light IMUs are not yet available. However, with weights of around 0.05 kg, large-scale production behind them, and the possibility to use redundant IMU setups we can expect significant progress towards DiSO precisions with 0.01 kg-level, INS/GNSS-based orientation systems in coming years.

Sensing payloads

UAS-based PaRS is a particular case of airborne PaRS and, as such, once the application requirements are set, the optimal combination of carrier—the UA—and sensing payload has to be found. In practice, it is often the case that the UA is predefined and, therefore, the UAS developer or operator faces the problem of fitting a remote sensing payload into the given UA volume, weight and power specifications, beside serving the specific application requirements (sensing bandwidth, accuracy, resolution, etc.) Finding the right balance is not only difficult but also complex, as the range of available systems is vast, ranging from low-cost mass-market, amateur and professional, to systems specifically designed for PaRS ones (including those designed for UAS and general systems).

A recent in-depth survey on optical remote sensing instruments, for terrestrial static, terrestrial kinematic, airborne and space image acquisition systems, can be found inRemondino (2011). Specifically for UAS, van Blyenburgh (2013) identifies 406 imaging and ranging instruments including active and passive systems, optical—from the visible band, to the Near Infrared (NIR) up to the Thermal Infrared (TIR) and microwave systems.

This section concentrates on remote sensing instruments which might be suitable for micro, mini and tactical UA payloads, particularly distinguishing between visible-band, near-infrared, multispectral, hyperspectral, thermal, laser scanners and synthetic aperture radar. Table 4, Table 5, Table 6, Table 7, Table 8 and Table 9 describe fundamental characteristics of some common and/or representative sensors, and aim to complement sensor integrators' knowledge on available systems.

Table 4. Common and/or representative small format (SF) and medium format (MF) visible band cameras.

Manufacturer and model	Format type	Resolution (MPx)	Size (mm^2)	Pixel size (μm)	Weight (kg)	Frame rate (fps)	Speed (s^{-1})
Phase One	MF	CCD	53.7	5.2	1.7	0.7	4000 (fp)
iXA 180		80	×40.4				1600 (ls)
Trimble	MF	CCD	53.7	5.2	1.5	–	1000 (ls)
IQ180		80	×40.4				
Hasselblad	MF	CCD	53.7	6	1.8	0.7	800 (ls)
H4D-60		60	×40.2				
Sony	SF	CMOS	23.5	3.9	0.35	2.3	4000 (fp)
NEX-7	MILC	24.3	×15.6				
Ricoh	SF	CMOS	23.6	4.8	0.35	3	3200 (fp)
GXR A16	IUC	16.2	×15.7				

fp: focal plane shutter, ls: leaf shutter.

Table 5. Common and/or representative multispectral cameras for UAS.

Manufacturer and model	Resolution (Mpx)	Size (mm^2)	Pixel size (μm)	Weight (kg)	Spectral range (nm)
Tertracam	CMOS	6.66	5.2	0.7	450–1050
MiniMCA-6	1.3	×5.32	×5.2		
Quest Innovations	CCD	10.2	7.5	0.8	400–1000
Condor-5 UAV-285	1.4	×8.3	×8.1		

Specifications on the MiniMCA-6 hold for each of the 6 sensors.

Table 6. Common and/or representative hyperspectral cameras for UAS.

Manufacturer and model	Resolution (Mpx)	Size (mm^2)	Pixel size (μm)	Weight (kg)	Spectral range (nm)	Spectral bands and resolution
Rikola Ltd.	CMOS	5.6	5.5	0.6	500–900	40
Hyperspectral Camera		×5.6				10 nm
Headwall Photonics	InGaAs	9.6	30	1.025	900–1700	62
Micro-Hyperspec X-series NIR		×9.6				12.9 nm

Table 7. Common and/or representative thermal cameras for UAS.

Manufacturer and model	Resolution (Mpx)	Size (mm^2)	Pixel size (μm)	Weight (kg)	Spectral range (μm)	Thermal sensitivity (mK)
FLIR	Uncooled VOx Microbolometer	10.8	17	0.07	7.5–13.5	⩽50
TAU 2 640	640 × 512	×8.7				
Thermoteknix Systems Ltd.	Amorphous Silicon	16	25	0.105	8–12	⩽50
Miricle 307K-25	640 × 480	×12.8				

Table 8. Common and/or representative laser scanners for UAS.

Manufacturer and model	Scanning pattern	Range (m)	Weight (kg)	Angular res. (deg)	FOV (deg)	Laser class and λ(nm)	Frequency (kp/s)	Application
ibeo Automotive Systems	4 Scanning	200	1	(H) 0.125	(H) 110	Class A	22	A
IBEO LUX	Parallel lines			(V) 0.8	(V) 3.2	905		
Velodyne	32 Laser/detector	100	2	(H) –	(H) 360	Class A	700	MM
HDL-32E	Pairs			(V) 1.33	(V) 41	905		
RIEGL	1 Scanning	⩾1000	–	(H) 0.01	(H) 60	Class 3B	200	H
VQ-820-GU	Line			(V) N/A	(V) N/A	532		

A: automotive; MM: terrestrial mobile mapping; H: hydrography.

Table 9. Common and/or representative synthetic aperture radars for UAS.

Manufacturer and model	Spectral bands	Weight (kg)	Transmitted power (W)	Resolution (m)
IMSAR NanoSAR B	X and Ku	1.58	1	between 0.3 and 5
Fraunhofer FHR MIRANDA	W	–	0.1	0.15
NASA JPL UAVSAR	L	200	2000	2
SELEX Galileo PicoSAR	X	10	–	1

NanoSAR B weight does not account for antenna and IMU.

Visible-band, near-infrared and multi-spectral cameras

The photogrammetric and remote sensing community have benefited from the mass-market and other professional markets' strength by leveraging these to design remote sensing instruments with high resolution.

Some developments consisting of multiple-head RGB cameras have been performed and results have been recently published. Xie et al. (2012) presents a wide-angle camera based on four single cameras Canon EOS 5D Mark II, including its calibration process and results. Grenzdörffer et al. (2012) describes the integration of five cameras (Crevis MV-CS27U USB) and its geometric and radiometric calibration procedures. Kohoutek and Eisenbeiss (2012) describes the use of a range-imaging (RIM) camera on a UAS, based on the time-of-flight (ToF) measurement principle, to measure distances to structures.

The maturation of visible-spectrum cameras has impacted several side-technologies, for example, mobile phones. Current smartphones come equipped with high-quality cameras at a reasonably low-cost. Yun et al. (2012) reports on a fixed-wing UAS equipped with Samsung Galaxy S and S2 smartphones to produce a digital elevation model (DEM) of a construction area in South Korea. More recently, the French company Lehman Aviation presented the LA300 UAS, featuring an on-board 41 megapixel Nokia Lumia 1020 optical camera.

Moving to the upper portion of the spectrum, several small, commercial multi-spectral cameras exist and have already been used in UAS for PaRS. This is the case of the mini-MCA series, from Tetracam, which is available in three configurations to capture four, six and twelve channels (Tetracam, 2012). They are able to discriminate spectral reflectance which is an important indicator in applications related to vegetation health. Bendig et al., 2012 and Lucieer et al., 2012 describe developments using mini-MCA cameras andKelcey and Lucieer (2012) reports on the geometric and radiometric calibration of a mini-MCA.

Table 4 summarizes common and/or representative RGB cameras suitable for UAS PaRS, as reported in several experiements. Bäumker and Przybilla (2011) report on the Ricoh GXR using the A16 lens, featuring 16.2 Mpx and variable focal length (24–85 mm), and Küng et al. (2011) report on the use of the Sony NEX-5. In the same product family, the Sony NEX-7 is present in commercial systems such as Falcon 8, from Ascending Technologies, or the Aeromapper, from Aeromao. In addition, some sensors are described as potential candidates in view of their specifications. For ease of comparison, we follow the same structure as in Remondino (2011) for the visible-band cameras.

Finally, Table 5 describes some of the most common developments in the field of multi-spectral sensing for UAS.

Hyperspectral cameras

Remote sensing with hyperspectral cameras deals with imaging narrow spectral bands over a continuous spectral range, producing the spectra of all pixels in the scene. In contrast, multispectral remote sensing produces discrete bands and usually has a lower spectral resolution. Hyperspectral sensors, therefore, extract more detailed information than multispectral sensors because an entire spectrum is acquired at each pixel. Contrary to the visible-spectrum camera developments, which have reached a weights of hundreds of grams and resolutions of tens of megapixels, the miniaturization process of multi- and hyper-spectral cameras is challenging in terms of optics and sensor calibration.

A combination of two commercial products has been announced for the first quarter of 2013, the fixed-wing UAS BRAMOR gHY, by C-Astral (Slovenia), and a small and light-weight hyper-spectral camera developed by Rikola Ltd. Rikola (2012).

In Rufino and Moccia (2005), a set of diverse sensors is presented in a UAS integration, consisting of a thermal camera, a three-band multispectral camera operating in the visible spectrum, and two hyperspectral sensors operating in the visible and NIR bands.

Table 6 provides information on recent relevant developments on hyperspectral sensors for UAS.

Thermal imaging

There has been noticeable progress in thermal imaging miniaturization in recent years. Low-weight, small-size imagers, such as those developed by FLIR, were first used in the military context for remote reconnaissance (Kostrzewa et al., 2003) and arc becoming more common in applications such as forest fire monitoring (Rufino and Moccia, 2005 and Scholtz et al., 2011). Aero-triangulation studies have also been carried out using thermal images (Hartmann et al., 2012).

Table 7 compiles some existing products in the family of thermal sensors, suitable for light UAS.

Laser scanners

While the use of laser scanners (or LiDAR, indistinctively) together with medium-large format cameras, is now common in traditional photogrammetry, their application to UAS for PaRS remains challenging, either due to the trade-off between performance and the size or cost of LiDAR, or the effect of flight dynamics on the measurement process (Zhou et al., 2012 and Wallace et al., 2012). Despite such difficulties, one of the first UA-borne LiDAR and camera integrations was presented a few years ago (Nagai et al., 2004). Other early integrations that followed them featured development for rapid mapping in emergency situations (Choi et al., 2009), and a compact and lightweight airborne laser scanner LMS-Q160 from Riegl mounted on a rotary wing Scout B1-100 (Imbach and Erk, 2009).

Short-range laser scanners (for example, from manufacturers such as FARO, SICK and Hokuyo) have already been integrated in UAS, but not for PaRS missions. In these cases, obstacle detection and avoidance (Scherer et al., 2008) or short-range mapping have been reported.

Table 8 presents recent LiDAR sensors which have been recently integrated with UAS. Recent integrations include the TerraLuma octocopter, which is equipped with an Ibeo LUX automotive LiDAR system (Wallace et al., 2012), or the premature 3D flash LiDAR described in Zhou et al. (2012), optimized in size and weight for UAS, presenting results for simulated flights at 300 m' altitude. In addition, the Velodynes HDL-32E has been successfully integrated on a Phoenix AL-2 multicopter

UAS, from Phoenix Aerial Systems, and the RIEGL VQ-820-GU hydrographic scanner, integrated in a Schiebel CAMCOPTER S-100.

Synthetic aperture radar

As for laser scanners, Synthetic Aperture Radar (SAR) technology has come a long way in traditional remote sensing. Yet its adaption to UAS (miniaturization versus performance) has still not been solved. However, there is some literature on the use of this technology, and few integration efforts into UAS have been reported world-wide. Indeed, the good performance of SAR despite adverse weather conditions is a very interesting feature not just for military groups, which have traditionally endorsed SAR technology, but also in PaRS.

Remy et al. (2012) describes a novel system for UAS, based on the combination of radar P- and X-bands for generation of digital terrain and elevation models in forested areas. InEssen et al. (2012), a development based on the W-band (millimeter-wave) is integrated within the NEO S300 helicopter-type UAS, to observe small-scale features and resolving up to 15 cm. In Schulz (2011), a platform development based on the SwissUAV NEO S-350 is presented along with the integration of a millimeter-wave SAR.

Table SAR compiles current SAR sensors integrated in UAS. The UAVSAR concept, developed by the NASA JPL (details in the table were extracted from Rosen et al. (2006)), is currently being integrated on Global Hawk UAS. More details on this integration were provided at the UAS Payloads Conference in Washington DC, June 19, 2012. The SELEX Galileo's PicoSAR has also been integragted into an Integrator UAS, from InSitu.

PROCESSING

As in traditional airborne PaRS, the processing of the data and measurements collected in a UAS mission is a key step in the development of UAS PaRS services. In principle, one would expect UAS PaRS to yield exactly the same type of products as airborne PaRS and that the corresponding production line paradigm—the "basic operations" (Dowman, 2012)—would be similar with some influences from close-range photogrammetry (Haala et al., 2011, Remondino et al., 2011 and Mayr, 2013). After all, it is a question of blocks of aerial images with some influences from close-range photogrammetry and also a question of close-range photogrammetry going aerial. Indeed, today, this is the case for a subset of the traditional PaRS products: orthophotos and elevation/surface models (assuming that the previous image orientation and camera calibration data are available). The tendency indicates that, in the near future, this will be the case for most of PaRS products since, once images are oriented and cameras calibrated, automatic, assisted or manual image interpretation (semantic feature extraction) to produce 3D models of various levels of detail can be conducted (Qin et al., 2013). (Deviation from parallelism of the image planes and possibly scale differences of overlapping images due to rotational and translational high frequency UA motion may hinder the stereoscopic capacity and visual comfort of human operators. However, convergent images can be rectified to the stereoscopic normal case and to a common scale.)

We note that the "old" photogrammetric community wishing to benefit from the next technological wave is reluctant to let a newcomer into the *santum sanctorum* of data processing without scrutinizing its performance and is prone to keep on using known software and methods "updated" for UAS (Cramer, 2013b and Mayr, 2013). This translates into more informative software outputs, more intermediate quality-control checks and more interactive editing tools. On the other hand, the "new" mapping community is less sensitive to and in lesser need of self-diagnosis tools and intermediate quality control checks and therefore more prone to the use of fully automated implementations of the same paradigm. There seems to be room for everyone, updated classics and newcomers.

On one hand, that UAS acqusition technologies, from platforms to sensors, are "lighter, smaller and simpler" does not necessarily translate into simpler processing software systems. In fact, in general, more sophisticated processing is required to compensate for the necessarily limited performance of small, lightweight platforms and acquisition systems. This situation is described well in Qin et al. (2013), a realistic paper where the challenges of processing—orientating and calibrating—a block of about 900 images are described. In a way, it could be said that photogrammetry and computer vision had to join forces to accurately and automatically process UAS-sourced images. On the other hand, "lighter, smaller and simpler" does not necessarily translate into second class results because the circumstances of UAS PaRS are different—and even more favorable—from those of traditional airborne PaRS: UAS PaRS can leverage the huge investment in mass-market navigation and imaging technologies, benefit from overall miniaturization trends and take advantage of its usually GSD.

Image orientation and camera calibration

An autopilot usually includes its own low-cost, light navigation system (NS). An orientation system (OS), depending on the orientation requirements, usually includes a mapping-grade or geodetic-grade set of sensors. In the former case, the NS time–Position–Velocity–Attitude (tPVA) solution can hardly be used as aerial control and therefore neither direct sensor orientation (DiSO) nor integrated sensor orientation (ISO) make much sense, and we have to go back to pure aerial triangulation or indirect sensor orientation (InSO). In this case, the orientation parameters provided by the NS simplify the automatic generation of tie points, their photogrammetric measurements and initial approximations to their coordinates. As a result, in practice, the orientation of unsorted sets of images by InSO is more of an academic exercise than a real issue in outdoor applications. In the latter case (Rehak et al., 2013) show that cm-level positioning can be achieved and both DiSO and ISO are possible.

Thus we arrive at the first question: InSO versus ISO. Advocates of a pure photogrammetric approach, that is InSO, argue that in small areas a dense set of Ground Control Points (GCPs) is easy and cheap to establish and, in general, obtain from existing orthophotomaps. Further, they claim that the fewer navigation and orientation sensors, the less MTOW and longer the autonomy. Those in favor of INS/GNSS (for their use in DiSO and ISO) argue that with less than 100 g (Table 3) cm- to dm-level positioning is feasible as proven by (Rehak et al., 2013). At the time this report was written, the vast majority of UAS imagery for mapping is being processed with the InSO method; i.e., deriving orientation and calibration parameters solely from photogrammetric measurements and GCPs. (In fact, while there are end-to-end commercial solutions for InSO, there are no comparable solutions for ISO yet.)

Within the InSO realm, there are three calibration sub-strategies. One option is to calibrate the camera shortly before or after the mission but previously to the bundle adjustment in a separate process as recommended in Remondino et al. (2011) where varying distances to object and convergent images guarantee the determinability of calibration parameters. We will refer to this as pre-calibration. (pre-calibration is also required for DiSO) A second option is to apply self-calibration as done, for instance, in the Pix4D software—with the Conrady-Brown (CB) (Brown, 1971) and interior orientation (IO) models—as reported in Cramer (2013b). The third option is to combine both, also as reported in Cramer (2013a) and suggested in Colomina et al. (2007). There are not many comparative analyses on the performance of the three strategies. In Cramer (2013b), both pre-calibration and self-calibration yield comparable results and, interestingly, the combination of pre-calibration and self-calibration does not bring any significant improvement. The risks of self-calibration with the CB and IO models in aerial bundle adjustment are known and well-illustrated in Vallet et al., 2011 and Rosnell and Honkavaara, 2012 where, for the same UAS image data set, rather different interior orientation parameters are recovered. On the other hand, the instability of some calibration parameters in the CB and IO calibration models explains why self-calibration with those parameters: (a) works well and that their value has a more contextual than absolute sense and (b) dominates over other self-calibration models.

Whatever the camera calibration strategy is, modern InSO is based on the automatic detection and image measurement of tie points. In traditional photogrammetry, this task has been solved since the beginning of the 1990s (Tsingas, 1992) and is known as automatic aerial triangulation (AAT). However, fifteen years later, the traditional, long established and proven photogrammetric AAT software was not able to process UAS blocks (Qin et al., 2013). In fact, there is nothing wrong with AAT. It was simply designed under completely different assumptions: interior orientation, radial and decentering distortions were assumed to be stable and therefore amenable for infrequent pre-calibration; block structure was assumed regular (almost nadir images, approximate constant scale, overlap and, within strips, attitude); and geometric and radiometric variations were known to be moderate.

On the contrary, the irregularity of UAS blocks was no obstacle to automatic image matching and bundle adjustment software originated in the computer vision community for more general purposes—the so-called Structure from Motion (SfM) approach (Snavely et al., 2008 and Agarwal et al., 2009),—or specifically for UAS flights (Küng et al., 2011), or in more recent photogrammetric AAT software like PhotoScan from AgiSoft as reported, for instance, in Gini et al. (2013). Computer vision techniques for automatic tie point generation are based on point detectors and descriptors of the SIFT type (Lowe, 2004) and its many variations or redesigns like SURF (Bay et al., 2008), ASIFT (Morel and Yu, 2009), BRIEF (Calonder et al., 2010) and LDAHash (Strecha et al., 2012). The tie point candidates are obtained image-wise with the point detectors. In a block of n images, in order to avoid the combinatorial computational explosion of $n \times (n-1)/2$ potential image overlaps, efficient algorithms, using more or less external information, have been derived to identify tie points in a reasonable time period. Some of these algorithms and related software packages have resorted to parallel processing; either through multi-core or through Graphical Processing Unit (GPU) computing.

Abdel-Wahab et al., 2012, Cramer, 2013b and Qin et al., 2013 show how to combine SfM and photogrammetric techniques and, in particular, initialize

Inpho's MATCH-AT and Leica's AAT with SfM tie points respectively. A similar approach is used in Rosnell and Honkavaara (2012) to initialize BAE's Socet Set software. Cramer (2013b) reports on the empirical point determination accuracy of two UAS flights with consumer-grade cameras, a Canon Ixus 100 IS and a Ricoh GXR Mount A12 with a Carl Zeiss Biogon objective, processed with a combined SfM and AAT approach. With it, horizontal accuracy is at the half GSD level ($\mu_E = 0.5$ GSD, $\mu_N = 0.4$ GSD) and vertical accuracy somewhat worse, $\mu_h = 0.7$ GSD. The same image data set processed with the Pix4D software (Strecha, 2011 and Strecha et al., 2012) yields a remarkable $\mu_E = \mu_N = 0.25$ GSD and $\mu_h = 1.0$ GSD.

An early review of the classical photogrammetric software packages in the context of UAS photogrammetry is given in Eisenbeiss (2009). From the five principal packages identified in Dowman (2012), at least two have announced new versions supporting UAS-sourced images. In contrast, photogrammetric software from younger companies like AgiSoft, SimActive, ICAROS, Menci Software, RACURS or Orbit GT seem to be better prepared to re-engineer their software rapidly and accept UAS-sourced images. Thus, Gini et al. (2013), for instance, compare the orientation and calibration performance of Pix4D with AgiSoft's PhotoScan and, for the latter, report empirical point determination accuracies of $\mu_E = 1.1$ GSD, $\mu_N = 0.4$ GSD and $\mu_h = 1.2$ GSD; or, in terms of flying height (FH), $\mu_E = 3.8 \times 10^{-4}$ FH, $\mu_N = 1.4 \times 10^{-4}$ FH and $\mu_h = 4.2 \times 10^{-4}$ FH. Remondino et al. (2012) analyze the performance of low-cost, free web service and open-source systems for the automatic orientation of UAS images in a close-range application (AgiSoft's PhotoScan, Microsoft's Photosynth, Microsoft and University of Washington's Bundler, IGN's APERO, and Google and University of Washington's VisualSfM). All software packages were able to deliver correct results as long as the network geometry was strong. However, they conclude that for large and complex data sets, the pure SfM approach is not free from reliability and repeatability problems. Note that APERO (Pierrot-Deseilligny and Cléry, 2011) is an open source tool that integrates computer vision and photogrammetric techniques.

Since INS/GNSS navigation in GNSS-denied or challenged environments may require the so-called "visual aiding"—i.e., the inclusion of tie features measurements in the navigation filter and the implicit or explicit use of the corresponding images' exterior orientation—we mention the related pioneering work in Grün (1985) on real-time bundle adjustment and the Simultaneous Localization and Mapping (SLAM) concept originated in the robotics community (Smith et al., 1986 and Leonard and Durrant Whyte,

1991). In a typical SLAM problem, measurements are modelled with stochastic differential equations (SDE) and stochastic equations (SE). The SDEs account for dynamic modelling e.g. the inertial mechanization equations that model the IMU measurements. Their integration yields the "prediction" step of the various SLAM estimation techniques. The SE—the "observation equations" in the geodetic language—are used to model measurements like GNSS ranges and image measurements, yielding the "update" step. The most popular related estimation technique of SLAM is the Kalman filter (Kalman, 1960), a closely related technique to sequential least-squares estimation. In some navigation problems, even in the absence of a dynamic model—i.e., an SDE,—the Kalman filter (KF) is used by the introduction of a trivial dynamic model like $\dot{x} = 0 + v$ with low weights. A recent departure from the classical KF known as incremental Smoothing and Mapping (iSAM) is given in Kaess et al. (2008), based on incremental bundle adjustment closely related to the work of Grün (1985). The literature of INS/GNSS, GNSS and INS navigation with optical sensor aiding for UAS is vast. The reader can consult the comprehensive survey in Kendoul (2012) or examples from the robotics (Dusha and Mejías, 2012) and photogrammetric community (Wang et al., 2008) respectively. Of particular interest for UAS applications is Visual SLAM where the primary instrument of navigation is an optical sensor. An interesting example of Visual SLAM is given in Strasdat et al. (2012) where the relevance of closing trajectory loops—i.e., tie and control features—is demonstrated.

Last, we note that, in the context of Kalman filtering and smoothing (KFS), a real-time solution (forward prediction-filtering step) is, in general, sub-optimal as compared to a post-processed solution (forward and backward prediction-filtering steps and final smoothing). Further, a real-time or even post-processed solution with KFS is, in general, sub-optimal as compared to a post-processed solution with least-squares network adjustment like a bundle adjustment because, in general, it cannot benefit from the tie point geometric constraints between strips.

Surface reconstruction

Digital surface models (DSM) and orthophotos are the two main mapping products of UAS PaRS. In principle, once images are oriented and, possibly, calibrated, deriving DSM and orthophotos are routine tasks that have already been automated in PaRS since more than 20 years ago (Krzystek, 1991). However, few questions (with still open answers) arise when dealing with UAS-based measurements, namely how conventional software for surface reconstruction perform with UAS images; in case of success, what is the quality (metric accuracy and morphological fidelity) of the obtained solutions; and what is the performance of the new generation software and techniques for DSM production. This section reviews literature regarding this issues.

Rosnell and Honkavaara (2012) investigated point cloud generation from UAS image sequences collected with two different camera setups, and studied the post-processing results using two software pieces, BAE systems' SocetSet, a traditional photogrammetric software, and Microsoft's Photosynth, an Internet-based service designed for photo-collection. Their conclusions indicate that aerial photogrammetric processing with the latter software is possible yet provides sparse point clouds and non-negligible distortions on the final surface model (vertical accuracy Root Mean Square (RMS) was estimated to oscillate between μ_h=40 GSD and μ_h=120 GSD). Moreover, the traditional post-processing software produced a point cloud almost as dense and accurate as the reference point cloud provided by a large-format photogrammetric camera (RMS height differences of μ_h=6 GSD). Yet, the authors observe problems (failed matching) in low altitude and large three-dimensional image sets, suggesting that conventional processing software might not still be flexible enough for UAS-like scenarios (close-range and oblique imagery).

Haala et al. (2013) present the results of processing aerial images from two different cameras using an in-house software development, named SURE. The paper assesses the empirical accuracy of the obtained DSM by measuring distances from points to planar patches extracted from a reference DSM. For a set of 33 patches extracted on two different areas, a Canon Ixus 100 IS yielded a vertical accuracy RMS of μ_h=0.5 GSD and μ_h=0.53 GSD, respectively, whilst a Ricoh GXR yielded a vertical accuracy RMS of μ_h=0.33 GSD and μ_h=0.28 GSD (the latter camera features a better signal-to-noise ratio due a bigger pixel size).

Harwin and Lucieer (2012) present a comprehensive compilation of the state-of-the-art techniques and their results for UAS-based point cloud generation, and present an accuracy study for the multi-view stereopsis technique. The authors estimate seven-parameter Helmert transformations for point cloud geo-referencing—this analysis goes beyond just one-dimensional height accuracy analysis. The study presents results for a set of different testing scenarios: variable GCP measurement methods (Total Station or Real-Time Kinematic GPS), different GCP distributions, number of GCPs used in the adjustment, or even types of GCP targets. As a summarizing result, the authors state that vertical accuracy between μ_h=2.5 GSD and μ_h=4 GSD can be achieved when flying between 40 and 50 m above ground, provided sufficient, clearly visible and evenly-distributed GCPs, and between 70% and 95% of overlap between images.

To our best knowledge there are no analyses concerning the morphological fidelity of elevation models obtained with UAS photogrammetry, an indication that contour lines are usually not derived from these type of point clouds.

In Fritz et al. (2013), a study is carried on to compare UAS-based point clouds using an frame camera and Terrestrial Laser Scanner (TLS) point clouds for tree stem reconstruction. No direct results on the point cloud generation are provided—rather, the study focuses tree detection and radius estimation. In these metrics, results point that reconstruction was less accurate and less dense than with TLS.

Besides frame cameras, LiDAR technology has played a major role in point cloud generation in conventional PaRS. Although extensively used for DSM generation via aerial or terrestrial acquisition, the use of LiDAR in UAS platforms has been (and still is) limited as already discussed in Section 5.3. Indeed, optical frame cameras have concentrated the focus of UAS commercial developments (see Table 10 as a compilation example), mainly pushed by a feasible miniaturization and cost-effectiveness of technology as already discussed. This circumstance, in combination with the integration of the computer vision research into UAS, has brought into scene state-of-the-art approaches for point cloud generation and/or densification using optical cameras in UAS, such as Structure-from-Motion (Hudzietz and Saripalli, 2011), Multi-View Stereopsis (Wefelscheid et al., 2011 and Harwin and Lucieer, 2012) and optimal flow (Pierrot-Deseilligny and Cléry, 2011). The last two methods are already implemented into the open-source packages PMVS and Micmac, respectively (Remondino et al., 2011).

Table 10. Examples of commonly used UAS in PaRS.

Name	Manufacturer	Weight (kg)	Endurance (h)	Integrated payload (i) or Payload weight (w)
Common fixed-wing unmanned aircraft				
SwingletCAM	SenseFly	0.5	0.5	(i) 16 Mpx RGB camera
GeoScan101	GeoScan	2	1	(i) 24.3 Mpx RGB camera
UX5	Trimble	2.5	0.83	(i) 16.1 Mpx MILC RGB camera
Pteryx	FotoMapy	5	2	(w) 1 kg w/o batteries
Sirius I	MAVinci	3	0.91	(i) 16 Mpx RGB camera
Kahu	Skycam	4	2	(i) Double-head 16 Mpx MILC RGB cameras
Common rotary-wing unmanned aircraft				
Geocopter	IGI	90	2	(w) 30 kg
Scout B1-100	Aeroscout	75	1.5	(w) 30 kg
R-MAX, type II	Yamaha	100	1	(w) 28 kg
Common multi-rotor unmanned aircraft				
md4-1000	Microdrones	3	1.46	1.2 kg
HT-8-2000	Height-Tech	2.4	0.28	2 kg
Aibot x6	Aibotix	2.4	30	2.5 kg
Falcon 8	Ascending technologies	1.45	0.33	0.75 kg

HexaKopter	MikroKopter	1.2	0.6	1 kg

Particularly, the introduction of Semi-Global Matching (SGM) (Hirschmüller, 2005) has been key in using optical cameras as a stand-alone solution for dense DSM production, and this is also reflected in UAS literature (Bulatov et al., 2011, Küng et al., 2011 and Haala et al., 2013). A comparative analysis of SGM and LiDAR is presented inGehrke et al. (2010) under the "controversial" question of whether SGM would take the lead over LiDAR for DSM generation. In this analysis, the SGM accuracy is described as "typically 0.5 GSD horizontally and 1.5 GSD vertically," and further testing with real images and LiDAR measurements show a vertical RMS difference of μ_h=1 GSD between SGM-based and LiDAR-based DSM, showing a remarkable coherence between the two products.

As opposed to old, comfortable times when manned flights shared similar patterns, the geometry of UAS-sourced photogrammetry is much more variable; and therefore, the extrapolation of the presented results to other UAS flights with different base-to-height ratios has to be made with caution.

UAS PaRS APPLICATIONS AND GEOMATIC MARKETS

The European Commission carried out a comprehensive study to monitor the uses of UAS in Europe, aiming to identify strengths and weaknesses in comparison with international developments (European Comission, 2007). In this study, a list of potential applications for civil and commercial UAS is provided consisting of *Non-Military Governmental* (Civil Security, Border Security, Coastguard); *Fire-fighting and Emergency Services* (forest fire spotting and co-ordination, major incident response co-ordination, emergency rescue), *Energy Sector and Communication Networks* (Oil and Gas industry distribution infrastructure, electricity grids, railway network monitoring),*Agricultural Forestry and Fisheries* (environmental monitoring, crop dusting, resource optimization); *Earth Observation and Remote sensing* (climate monitoring, aerial photography, mapping and surveying, seismic monitoring, pollution monitoring), and *Communications and Broadcasting* (VHALE platforms as proxy-satellites, MALE UAS for communication coverage, camera platforms). Modulated to the scope of our paper, this chapter provides a review of the state-of-the-art initiatives regarding just a few of the applications featured in the previous list (the names of the applications may be also modified for convenience).

Agricultural and environmental applications

Remote sensing is a well-known art for agriculture and environment analysis. Vegetation and/or biodiversity control has been traditionally performed using aerial and/or satellite imagery, resulting in high expenses when fine resolution is requested. UAS have successfully introduced the *smaller, cheaper-to-operate* platform paradigm among the remote-sensing community. The range of available sensors is widening as a natural attempt to adapt to smaller platforms, in which weight and dimension restrictions hold as opposite to manned aerial platforms, and also to adapt to user and application needs.

Several researchers and/or companies have balanced the requirements of the payload and aerial platform to enable the operation of small, fast and easily deployable systems and cover small or medium-size areas. For example, Rufino and Moccia (2005) used a radio-controlled fixed-wing model to fly a thermal imager and a hyperspectral sensor in visible-NIR bands targeting forest fire monitoring. Another example is Zarco-Tejada and Berni (2012), in which a miniaturized hyperspectral camera mounted on a fixed-wing auto-piloted platform of 6 kg MTOW is described. In Bendig et al. (2012) a mini-UAS MK-Okto by HiSystems GmbH equipped with either a NEC F30 IS thermal imaging system or a tetracam Mini MCA-4 is described for successful Normalized Difference Vegetation Index (NDVI) computation. Gini et al. (2012) describes the use of a Pentax Optio A40 for RGB photos and a Sigma DP1 modified to acquire the NIR band, on-board a Microdrones md4-200, for tree classification based on different vegetation indices.Agüera et al. (2011) describes the use of the same platform, equipped with an ADC Lite Tetracam, to compare aerial and ground measurements and vegetation indices. Lucieer et al. (2012) uses an Oktokopter with optical and hyperspectral cameras to analyze the Antarctic moss beds, and Jensen et al. (2012) presents an RGB, NIR and thermal-vision concept to monitor stream termperatures. Costa et al. (2012) presents a combination of UAS and a ground wireless sensor network to proceed with crop fertilizing missions. In it, the UAS route is modified depending on the inputs from the ground network, which can measure the amount of fertilizer applied. In Grenzdörffer and Niemeyer (2011), the use of UAS for Bidirectional Reflectance Distribution Function (BRDF) measurements is proposed as an alternative to costly and cumbersome field goniometer measurement campaigns, in the context of agricultural applications.

An example of environmental application is presented in Wich and Koh (2012), in which small fixed-wing UAS carrying photo- or video-cameras are used in missions in Switzerland, the Netherlands, Indonesia, Malaysia and Nepal, to perform detection of several species such as orangutans, elephants or rhinos and provide information on density and circulation of animals. The use of small UAS to detect animals is also the motivation behind (Israel, 2011), in which a thermal camera Tau640 from FLIR is flown on-board a Falcon 8 from Ascending Technologies GmbH to detect small fawns in meadows before mowing machines could harm them. Grenzdörffer (2013) reports on the use of UAS for bird counts.

In the geological field, Eisenbeiss (2009) reports on the use of a rotary-wing Copter 1B equipped with a Nikon D2Xs to perform oblique-view mapping of the mountainous area of Randa, Switzerland. Its mission is to analyze tectonic fractures. In Delacourt et al. (2009), a coastal management application, related to the quantification of morphosedimentary changes of the coastal fringe, is carried out with a rotary-wing platform carrying a commercial digital reflex camera to generate a DEM for hydrodynamics numerical modelling. Eck and Imbach (2011) reports on the use of a high-resolution 3-axis magnetic sensor, mounted on an autonomous Scout B1-100 helicopter to generate detailed magnetic maps.

Cox et al. (2006) is a report performed in 2006 by the NASA's Civil UAV Team, in which a list of NASA-funded science mission experiences since 1995 are compiled, such as clear air radiation measurements, cumulus electrification measurements, harvest optimization, coastal mapping, atmospheric chemistry and many others. It also provides an assessment of Earth science, Land Management and Homeland versus their required capabilities (access to regulated airspace, long endurance, quick deployment, etc.).

Intelligence, surveillance, and reconnaissance

Aerial observation by unmanned platforms has been powered basically in the military context. Indeed, small, especially handheld or hand launched, UAS can provide an "over-the-hill" point of view to an army's ground troops, in order to avoid unseen potiential dangers. In larger UAS categories, large unmanned aircraft provide broad-area surveillance i.e. border control and restricted area surveillance. UAS have also been used as communications relays to increase battlefield awareness, or as decoys to fool enemy's radars. Close to the military segment, search-and-rescue or disaster management missions share many objectives with the previously specified (basically, providing quick imagery from an area where no supporting structures can be assumed), and thus can somehow be classified as Intelligence, Surveillance, and Reconnaissance (ISR) missions. We note that tactical UAS (or bigger) is usually preferred for ISR as they implement real-time image or video downloads easier than mini or micro UAS.

As illustrative examples of military-related UAS, two developments are described. Insitu, a Boeing company, has developed the ScanEagle, a 20 kg MTOW fixed-wing with a wingspan of 3 m, used by the U.S. Navy and Marine Corps. This system delivers imagery enabling tactical commanders to develop a clearer picture of the battlefield. As standard payload, it carries either an inertially stabilized electro-optical or an infrared camera. The gimbaled camera allows the operator to easily track both stationary and moving targets, providing real-time intelligence. Another development by Insitu is the Integrator, which has approximately the same dimensions but of larger MTOW, as it is designed to provide high payload capacity and modularity to include different sensors for each mission. The Integrators baseline sensor package includes inertially stabilized electro-optic, long-wave infrared and mid-wave infrared cameras, with infrared marker and laser rangefinder.

Hereafter, some scientific work is compiled in relation the ISR UAS. Molina et al. (2012)report on the use of a rotary-wing UA equipped with video and thermal cameras to detect lost persons in difficult-to-access situations, that is, to add value in the search component of search-and-rescue, and (van Persie et al., 2011) describes a similar development, using a video camera on a rotary-wing platform, to support fire brigades in real-time crisis management. Another rotary-wing development is presented in Choi and Lee (2011) for rapid disaster management, carrying two optical cameras and a LiDAR, and presenting results including the geo-referenced LiDAR point cloud.

Aerial monitoring in engineering

Again, the "above-the-head" privileged point of view that UAS provide is the main motivation for using them in civil engineering or, in general, in any engineering requiring infrastructure monitoring. Some of the actual infrastructures of interest for inspection are high and medium voltage lines, oil and gas pipe lines, roads, railways, etc.

Merz and Chapman (2011) presents the design of a helicopter-type UAS targeting infrastructure inspections and crop monitoring missions using an RGB camera and convenient filters. The use of a Commercial-Off-The-Shelf (COTS) 2D LiDAR on-board to enable terrain-based navigation, and thus to ensure beyond line-of-sight operation in a priori unknown areas, is noteworthy.

When combined with the suitable remote sensing tools, UAS can be also powerful resources for energy efficiency and wealth management. Jensen et al. (2009) describes an approach using multiple swinglet-type UAS to perform distributed wind measurement, which is of interest for meteorologists and also wind farms. Matsuoka et al. (2012) reports on an experiment conducted in order to investigate the feasibility of the deformation measurement of a large-scale solar power plant on reclaimed land by using images acquired by a non-metric digital camera on board a micro UAS.

Ground monitoring is a common application in remote sensing, for which ground-based and satellite-based tools are widely used, and recent studies show that UAS have potential for this application. Rau et al. (2011) presents the use of a fixed-wing platform with a consumer-grade camera to perform landslide detection and vegetation indices computation. Niethammer et al. (2011) studies landslide monitoring through the generation of ortho-mosaic and DTMs with the use of open-source software. Carvajal et al. (2011) describes the use of a quadracopter carrying a 12 Mpx camera to characterize landslides on road ditches. Shi et al. (2011) presents a method for object-change detection dealing with large rotations in pairs of UAS-based images, aiming at infrastructure monitoring applications.

Cultural heritage

The surveying of archaeological sites using UAS has now become common. Again, the ease of UAS operation has been key for its choice and, also very important, the quality of processed measurements has reached a level sufficient to convince the cultural heritage community. The topic is now well-established in most of the PaRS congresses (see UAV-g congress, in editions 2011 and 2013).

Rinaudo et al. (2012) describes the use of a Hexakopter, by Mikrokopter, equipped with a Sony NEX-5 to generate the DSM and orthophoto of a Roman villa archaeological site located in Aquileia (Italy), a well-known UNESCO WHL site. In Seitz and Altenbach (2011), a quadracopter by the same company is used, together with a 14 Mpx camera, to produce image mosaics and 3D models in Germany and Cambodia. Mészáros (2011)reports on the ortho-mosaic production in a recently discovered ruin in Hungary using a fixed-wing UAS, an RGB camera and an in-house open-source autopilot.

Remondino et al. (2011) describes the aerial image acquisition campaign carried out in Veio, Italy, over the ancient Etruscan city. Using Microdrone quadri-rotors equipped with a Pentax Optio A40, aerial images with 1 cm GSD were acquired and a dense matching using MicMac post-processing software delivered a point cloud of around 40 million points. Another documented area of study is the Maya site in Copan, Honduras, over which a model helicopter equipped with a Nikon SRL camera of 12 Mpx was flown, obtaining GSD of 1 cm and delivering DSM. Finally, a heritage area in Pava, Italy was also flown with a Microdrone MD4-200, again with 1 cm GSD and producing DSM at 5 cm in resolution. Comparison with check points show around 3 and 2 cm in planimetry and height, respectively. Additionally, this paper includes a structured analysis and review of UAS-based PaRS, from platforms to processing and looking forward to future perspectives.

Research has been performed on the combination of TLS and oblique aerial imagery from UAS (Fiorillo et al., 2012 and Eisenbeiss, 2009) for 3D reconstruction which is often the preferred product to be delivered in archaeology and ancient building surveying.

Traditional surveying, conventional mapping and photogrammetry, and cadastral applications

Cramer et al. (2013) provides a state-of-the-art overview on the use of UAS by some European National Mapping Agencies (NMAs). The consulted NMAs were aware of the potential of UAS. They follow most recent developments and work on a possible integration of UAS data in their production lines. Many of those NMAs are also involved in local uses, such as cadastral applications, land management/land consolidation or disaster monitoring.

Manyoky et al. (2011) describes the use of a UAS—a Falcon 8 with a Panasonic Lumix DMC-LX3—, for cadastral surveying and its comparison with conventional data acquisition methods. Further, Cunningham et al. (2011) analyses the opportunities and presents the experiences on the use of UAS for cadastral mapping, focusing on rural Alaska.

van Hinsberg et al. (2013) report on the use of UAS for high-precision parcel boundary determination, of up to 3 cm which is similar to conventional land surveying. This exercise is important when ownership of one or more parts of a parcel changes, as seller(s) and buyer(s) are legally obliged to identify the new boundaries and, usually, are unable to attend the on-site identification session.

Eyndt and Volkmann (2013) claim that UAS PaRS can be considered as another surveying tool, and even that, in several situations, it is an advantageous alternative to traditional surveying.

Mayr, 2011a and Gülch, 2012 describe the Personal Aerial Mapping System (PAMS) system, from SmartPlanes (Sweden), composed of a small fixed-wing platforms and an RGB camera, for DTM/DSM and orthomosaic generation. The former paper is a user report, that describes first-hand experiences in operating the system, and the later paper reports on the application of standard photogrammetric pipeline for the processing of UAS images, using software packages as MATCH-AT, MATCH-T DSM, OrthoMaster and OrthoVista.

Harwin and Lucieer (2012) proposes the application of the multi-view stereopsis technique, combining photogrammetry and computer vision, to imagery acquired from a multi-rotor micro-UAS of a natural coastal site in southeastern Tasmania, Australia. The point cloud is further densified with patch-based multi-view stereo techniques, producing 1–3 cm point spacing, and conclusions point that sub-decimeter coastal erosion can be monitored.

Grün et al. (2012) presents the use of high-resolution satellite imagery (GeoEye-1) to produce DTM, and its completion by using UAS imagery to produce high resolution 3D models of man-made structures, applied in this case to historical buildings in Bhutan.

Qin et al. (2013) reports on a mission in a quite complex urban area in the tropical city of Singapore with a Falcon 8 octocopter, developed by Ascending Technologies GmbH, with an off-the-shelf Sony NEX-5 camera.

CONCLUSIONS

We have reviewed the UAS technology for PaRS applications with emphasis on regulations, acquisition systems, navigation and orientation. The diversity and sophistication of the involved technologies is apparent: aeronautics, satellite and inertial navigation, computer vision, robotics, sensorics and, last not least, photogrammetry. Technologically speaking, UAS-sourced PaRS are mature enough to support the development of geoinformation products and services. At least for a first generation operational tools. Moreover, in spite of a still emerging and uncertain regulatory frame, customer demand and general interest is present to the point that there are already some UAS geoinformation niche markets; in particular a growing new market for small photogrammetric and remote sensing projects. The trend seems to be unstoppable. The majority of commercial PaRS applications are conducted with micro UAS and with off-the-shelf cameras. For this type of equipment, flight planning, flight navigation, guidance and control and data post-processing software exist. In particular, PaRS post-processing software for sensor orientation/calibration and for surface reconstruction is commercially available and production capable. It is already developed to a high level of automation to serve a market whose workforce is, in general, less specialized than that of the traditional airborne and space market. Computer vision techniques—originating from the computer vision community or from the modern PaRS community—have been instrumental in achieving this level of automation. INS/GNSS trajectory parameters, though necessary for real-time UA navigation and control, are seldom used in the mentioned post-processing orientation/calibration software—for DiSO and ISO—because of the performance limitations of current miniature IMUs. However, the pace of development of MEMS tactical-grade IMUs can change the orientation/calibration software landscape rapidly. Subsequent surface reconstruction and orthophoto generation leverage the recent achievements of high-density matching and the investments in the design of amateur and mass-market cameras. All in all, for small projects, UAS PaRS offers an unbeatable price-performant service and product. UAS PaRS is one of the rapidly developing UAS market niches like agriculture and others. Were it not because of them, we would not have the big regional players preparing UAS integration and regulation plans (ERSG and FAA roadmaps for the intergation of UAS in the airspace) and UAS testbeds (FAA). At the other end, small countries (recall the Trinidad and Tobago public bid for a national UAS-sourced mapping project) and regions plan UAS-sourced photo coverages. Additional research and technology are clearly required and are being conducted with the advantage that the entrance barriers in the field are low. We will definitely see more new UAS technologies and applications in PaRS in coming years.

APPENDIX A. UAS PICTURES AND COMMON PARS UAS

In this annex, Fig. 1 presents some pictures from various systems, corresponding to each of the categories provided in van Blyenburgh (2013). All photos have been extracted from the UVS International Photo

Library, supplied by UVS International, and are protected by the copyright of the producing company. The authors would like to thank UVS International for the supplied materials.

Figure 1. From top to bottom, left to right, each picture illustrates a UAV of each category in van Blyenburgh (2013): AeroVironment, USA—Nano-Hummingbird; Ascending Technologies GmbH, Germany—Falcon 8; CATUAV, Spain—Argos; Swiss UAV, Switzerland—Neo s300; Schiebel, Austria—Camcopter S100; MMIST, Canada—Snowgoose; Thales, UK—Watchkeeper; Selex ES, Italy—Nibbio; Insitu Inc., USA—Integrator; General Atomics Aeronautical Systems, USA—Predator A; QinetiQ, UK—Zephyr; Lockheed Martin, USA—Morphing UAS.

In addition, Table 10 presents UAS of three different types (fixed-wing, rotary-wing and multi-rotor) commonly used among the PaRS community, and their main characteristics are described.

REFERENCES

1. Abdel-Wahab, M., Wenzel, K., Fritsch, D., 2012. Efficient reconstruction of large unordered image datasets for high accuracy photogrammetric applications. ISPRS – Ann. Photogramm. Remote Sens. Spatial Inform. Sci. I-3, 1–6.
2. Federal Aviation Administration, 2013. Integration of civil Unmanned Aircraft Systems (UAS) in the National Airspace System (NAS) Roadmap. Technical Report. US Department of Transportation, Federal Aviation Administration. Washington, DC, USA.
3. Agarwal, S., Snavely, N., Simon, I., Seitz, S.M., Szeliski, R., 2009. Building Rome in a day. In: 2009 IEEE 12th International Conference on Computer Vision, Sept. 29 2009-Oct. 2 2009, pp.72–79. http://dx.doi.org/10.1109/ICCV.2009.5459148.
4. Agüera, F., Carvajal, F., Pérez, M., 2011. Measuring sunflower nitrogen status from an unmanned aerial vehicle-based system and an on the ground device. ISPRS –Ann. Photogramm. Remote Sens. Spatial Inform. Sci. XXXVIII-1/C22, 33–37.
5. Barnard, J., 2007. Small UAV (<150 kg TOW) Command, Control and Communication Issues. Technical Report. Institution of Engineering and Technology.
6. Bäumker, M., Przybilla, H., 2011. Investigations on the accuracy of the navigation data of unmanned aerial vehicles using the example of the system Mikrokopter. ISPRS – Ann. Photogramm. Remote Sens. Spatial Inform. Sci. XXXVIII-1/C22, 113–118.
7. Bay, H., Ess, A., Tuytelaars, T., van Gool, L., 2008. Speeded-up robust features (SURF). Comput. Vision Image Understand. 110, 346–359, Similarity Matching in Computer Vision and Multimedia.
8. Bendig, J., Bolten, A., Bareth, G., 2012. Introducing a low-cost mini-UAV for thermaland
9. multispectral-imaging. ISPRS – Int. Arch. Photogramm. Remote Sens. Spatial Inform. Sci. XXXIX-B1, 345–349.
10. Blázquez, M., Colomina, I., 2012a. Fast AT: a simple procedure for quasi direct orientation. ISPRS J. Photogramm. Remote Sens. 71, 1–11.

11. Blázquez, M., Colomina, I., 2012b. Relative INS/GNSS aerial control in integrated sensor orientation: models and performance. ISPRS J. Photogramm. Remote Sens. 67, 120–133.
12. Brissset, P., Drouin, A., 2004. PaparaDzIY: do-it-yourself UAV. J. Micro Drones, Toulouse, FR.
13. Brown, D., 1971. Close range camera calibration. Photogramm. Eng. 37, 855–866.
14. Bulatov, D., Solbrig, P., Gross, H., Wernerus, P., Repasi, E., Heipke, C., 2011. Contextbased urban terrain reconstruction from UAV-videos for geoinformation application. ISPRS – Int. Arch. Photogramm. Remote Sens. Spatial Inform. Sci. XXXVIII-1/C22, 75–80.
15. CAA, 2012a. Air Navigation: The Order and the Regulations. CAP 393. Civil Aviation Authority (UK), London, UK.
16. CAA, 2012b. Unmanned Aircraft System Operations in UK Airspace – Guidance. CAP 722. Civil Aviation Authority (UK), London, UK.
17. Calonder, M., Lepetit, V., Strecha, C., Fua, P., 2010. BRIEF: binary robust independent elementary features. In: Daniilidis, K., Maragos, P., Paragios, N. (Eds.), Computer Vision—ECCV 2010, Lecture Notes in Computer Science, vol. 6314. Springer, Berlin Heidelberg, pp. 778–792.
18. Carvajal, F., Agüera, F., Pérez, M., 2011. Surveying a landslide in a road embankment using unmanned aerial vehicle photogrammetry. ISPRS – Int. Arch. Photogramm. Remote Sens. Spatial Inform. Sci. XXXVIII-1/C22, 201–206.
19. CASA, 2002. Unmanned Aircraft and Rockets: Model Aircraft. Advisory Circular AC- 101-3(0). Civil Aviation Safety Authority Australia. Canberra, Australia.
20. Chao, H., Cao, Y., Chen, Y., 2010. Autopilots for small unmanned aerial vehicles: a survey. Int. J. Control Autom. Syst. 8, 36–44.
21. Cho, G., Hildebrand, A., Claussen, J., Cosyn, P., Morris, S., 2013. Pilotless aerial vehicle systems: size, scale and functions. Coordinates 9, 8–16.
22. Choi, F., Lee, I., 2011. A UAV-based close-range rapid aerial monitoring system for emergency responses. ISPRS – Int. Arch. Photogramm. Remote Sens. Spatial Inform. Sci. XXXVIII-1/C22, 247–252.
23. Choi, K., Lee, I., Hong, J., Oh, T., Shin, S.W., 2009. Developing a UAV-based rapid mapping system for emergency response. The International Society for Optical Engineering (SPIE) Proceedings 7332, 9–12. http://dx.doi.org/10.1117/ 12.818492.

24. Colomina, I., Aigner, E., Agea, A., Pereira, M., Vitoria, T., Jarauta, R., Pascual, J., Ventura, J., Sastre, J., Brechbühler de Pinho, G., Derani, A., Hasegawa, J., 2007. The uVISION project for helicopter-UAV photogrammetry and remote-sensing. In: Proceedings of the VII International Geomatic Week, Institute of Geomatics, Barcelona, Spain.
25. Colomina, I., Blázquez, M., Molina, P., Parés, M., Wis, M., 2008. Towards a new paradigm for high-resolution low-cost photogrammetry and remote sensing. ISPRS – Int. Arch. Photogramm. Remote Sens. Spatial Inform. Sci. XXXVII-B1, 1201–1206.
26. European Comission, 2007. Study Analysing the Current Activities in the Field of UAV. Technical Report. ENTR/2007/065, Second Element: Way Forward.
27. Cosentino, F., 2008. Development of an Integral System for Aerophotogrammetric Flight Planning, Data Acquisition and Imagery Ortho-mosaicking, Aerodreams UAV, accessed on February 2014.
28. Costa, F.G., Ueyama, J., Braun, T., Pessin, G., Osorio, F.S., Vargas, P.A., 2012. The use of unmanned aerial vehicles and wireless sensor network in agricultural applications. In: Geoscience and Remote Sensing Symposium (IGARSS), 2012 IEEE International, 22-27 July, pp. 5045–5048. http://dx.doi.org/10.1109/IGARSS.2012.6352477.
29. Cox, T., Somers, I., Fratello, S., 2006. Earth Observation and Role of UAVs: A Capabilities Assessment, Version 1.1. Technical Report. Civil UAV Team, NASA.
30. Cramer, M., 2013a. RPAS im Einsatz fr die Datenerfassung beim LGL BW. In: Presentation slides from the UAV Dach meeting, held in Rostock, Germany,2013.
31. Cramer, M., 2013b. The UAV LGL BW project—a NMCA case study. In: 54th Photogrammetric Week. Institut für Photogrammetrie, Universität Stuttgart, pp.165–179.
32. Cramer, M., Bovet, S., Gültlinger, M., Honkavaara, E., McGill, A., Rijsdijk, M., Tabor, M., Tournadre, V., 2013. On the use of RPAS in national mapping—the EuroSDR point of view. ISPRS – Int. Arch. Photogramm. Remote Sens. Spatial Inform. Sci. XL-1/W2, 93–99.
33. Cunningham, K., Walker, G., Stahlke, E., Wilson, R., 2011. Cadastral audit and assessments using unmanned aerial systems. ISPRS – Int. Arch. Photogramm. Remote Sens. Spatial Inform. Sci. XXXVIII-1/C2, 213–216.
34. Dalamagkidis, K., Valavanis, K., Piegl, L., 2009. On integrating unmanned aircraft systems into the national airspace system: issues, challenges, operational restrictions, certification, and recommendations. Intelligent Systems, Control and Automation: Science and Engineering, vol. 36. Springer-Verlag.

35. Davis, B., 2013. FAA Approves Test Sites to be ''Broad Platform for UAS Operations. Unmanned Systems News 30/12/2013. Technical Report. Association for Unmanned Vehicle Systems International AUVSI.
36. Delacourt, C., Allemand, P., Jaud, M., Grandjean, P., Deschamps, A., Ammann, J., Cuq, V., Suanez, S., 2009. DRELIO: An unmanned helicopter for imaging coastal areas. J. Coast. Res. 2, 1489–1493.
37. van Diggelen, F., 2010. Are we there yet? The state of the consumer industry. GPSWorld 21, 8–10, 48.
38. van Diggelen, F., Abraham, C., de Salas, J., Silva, R., 2011. GNSS inside mobile phones: GPS, GLONASS, QZSS, and SBAS in a single chip. InsideGNSS 6, 50–60.
39. Dowman, I., 2012. More than information from imagery. Geospatial World 2, 34–44.
40. Dusha, D., Mejías, L., 2012. Error analysis and attitude observability of a monocular GPS/visual odometry integrated navigation filter. Int. J. Robot. Res. 31, 714–737.
41. Dusza, B., Wietfeld, C., 2010. Performance evaluation of IEEE 802.16e mobile WiMAX for long distance control of UAV swarms. In: 2010 IEEE International Conference on Wireless Information Technology and Systems (ICWITS), Aug. 28 2010–Sept. 3 2010, pp. 1–4. http://dx.doi.org/10.1109/ICWITS.2010.5611937.
42. Eck, C., Imbach, B., 2011. Aerial magnetic sensing with an UAV helicopter. ISPRS – Int. Arch. Photogramm. Remote Sens. Spatial Inform. Sci. XXXVIII-1/C22, 81–85.
43. Eisenbeiss, H., 2009. UAV Photogrammetry. Ph.D. Thesis. Institut für Geodesie und Photogrammetrie, ETH-Zürich. Zürich, Switzerland.
44. Elkaim, G., Lie, F., Gebre-Egziabher, D., 2014. Principles of guidance, navigation and control of UAVs. In: Valavanis, K., Vachtsevanos, G. (Eds.), Handbook of Unmanned Aerial Vehicles. Springer, p. 1600.
45. ERSG, 2013. Roadmap for the integration of civil Remotely-Piloted Aircraft Systems into the European Aviation Systems. Technical Report. European RPAS Steering Group.
46. Essen, H., Johannes, W., Stanko, S., Sommer, R., Wahlen, A., Wilcke, J., 2012. High resolution W-band UAV SAR. In: 2012 IEEE International Geoscience and Remote Sensing Symposium (IGARSS), 22-27 July 2012, pp. 5033–5036. http:// dx.doi.org/10.1109/IGARSS.2012.6352480.
47. Everaerts, J., 2009. NEWPLATFORMS – Unconventional Platforms (Unmanned Aircraft Systems) for Remote Sensing. Technical Report 56. European Spatial Data Research (EuroSDR).

48. Everaerts, J., Lewyckyj, N., 2011. Obtaining a permit-to-fly for a HALE-UAV in Belgium. ISPRS – Int. Arch. Photogramm. Remote Sens. Spatial Inform. Sci. XXXVIII-1/C22, 1–5.
49. Everaerts, J., Lewyckyj, N., Fransaer, D., 2004. PEGASUS: Design of a stratospheric long endurance UAV system for reomte sensing. ISPRS – Int. Arch. Photogramm. Remote Sens. XXXV-B2, 29–33.
50. Eyndt, T., Volkmann, W., 2013. UAS as a tool for surveyors: from tripods and trucks to virtual surveying. GIM Int. 27, 20–25.
51. Fiorillo, F., Jiménez, B., Remondino, F., Barba, S., 2012. 3D surveying and modeling of the archaeological area of Paestum, Italy. Virtual Archaeol. Rev. 4, 55–60.
52. Fritz, A., Kattenborn, T., Koch, B., 2013. UAV-based photogrammetric point clouds— tree stem mapping in open stands in comparison to terrestrial laser scanner point clouds. ISPRS – Int. Arch. Photogramm. Remote Sens. Spatial Inform. Sci.,141–146.
53. Gakstatter, E., 2010. What about when high accuracy is cheap? GPSWorld 21, 16.
54. Gehrke, S., Morin, K., Downey, M., Boehrer, N., Fuchs, T., 2010. Semi-global matching: an alternative to lidar for dsm generation? Symp. Commiss. I, ISPRS 1, 2.
55. Gini, R., Pagliari, D., Passoni, D., Pinto, L., Sona, G., Dosso, P., 2013. UAV photogrammetry: block triangulation comparisons. ISPRS – Int. Arch. Photogramm. Remote Sens. Spatial Inform. Sci., 157–162.
56. Gini, R., Passoni, D., Pinto, L., Sona, G., 2012. Aerial images from a UAV system: 3D modelling and tree species classification in a park area. ISPRS – Int. Arch. Photogramm. Remote Sens. Spatial Inform. Sci. XXXIX-B1, 361–366.
57. Goodall, C., Carmichael, S., El-Sheimy, N., Scannell, B., 2012. INS face off MEMS versus FOGs. InsideGNSS 7, 48–55.
58. GORTT, 2013. Supply and Installation of an Unmanned Aerial Vehicle Mapping System and Terrestrial Laser Scanner for the Division of Lands and Surveys.
59. Technical Report. Ministry of Housing, Government of the Republic of Trinidad and Tobago (GORTT). Port of Spain, Trinidad and Tobago.
60. Grenzdörffer, G., Niemeyer, F., 2011. UAV-based BRDF-measurements of agricutltural surfaces with PFIFFikus. ISPRS – Int. Arch. Photogramm. Remote Sens. Spatial Inform. Sci. XXXVIII-1/C22, 229–234.
61. Grenzdörffer, G., Niemeyer, F., Schmidt, F., 2012. Development of four vision camera system for a micro-UAV. ISPRS – Int. Arch. Photogramm. Remote Sens. Spatial Inform. Sci. XXXIX-B1, 369–374.

62. Grenzdörffer, G.J., 2013. UAS-based automatic bird count of a common gull colony. ISPRS – Int. Arch. Photogramm. Remote Sens. Spatial Inform. Sci. XL-1/W2, 169–174.
63. Grün, A., 1985. Algorithmic aspects in on-line triangulation. Photogramm. Eng. Remote Sens. 51, 419–436.
64. Grün, A., Zhang, Z., Eisenbeiss, H., 2012. UAV photogrammetry in remote areas – 3D modeling of Drapham Dzong, Bhutan. ISPRS – Int. Arch. Photogramm. Remote Sens. Spatial Inform. Sci. XXXIX-B1, 375–379.
65. Gülch, E., 2012. Photogrammetric measurements in fixed wign UAV imagery. ISPRS – Int. Arch. Photogramm. Remote Sens. Spatial Inform. Sci. XXXIX-B1, 381–386.
66. Haala, N., Cramer, M., Rothermel, M., 2013. Quality of 3D point clouds from highly overlapping UAV imagery. ISPRS – Int. Arch. Photogramm. Remote Sens. Spatial Inform. Sci. XL-1/W2, 183–188.
67. Haala, N., Cramer, M., Weimer, F., Trittler, M., 2011. Performance test on UAV-based photogrammetric data collection. ISPRS – Int. Arch. Photogramm. Remote Sens. Spatial Inform. Sci. XXXVIII-1/C22, 7–12.
68. Haarbrink, R., 2011. UAS for geo-information: current status and perspectives. ISPRS – Int. Arch. Photogramm. Remote Sens. Spatial Inform. Sci. XXXVIII-1/C22, 207– 212.
69. Hartmann, W., Tilch, S., Eisenbeiss, H., Schindler, K., 2012. Determination of the UAV position by automatic processing of thermal images. ISPRS – Int. Arch. Photogramm. Remote Sens. Spatial Inform. Sci. XXXIX-B6, 111–116.
70. Harwin, S., Lucieer, A., 2012. Assessing the accuracy of georeferenced point clouds produced via multi-view stereopsis from unmanned aerial vehicle (UAV) imagery. Remote Sens. 4, 1573–1599.
71. van Hinsberg, W., Rijsdijk, M., Witteveen, W., 2013. UAS for cadastral applications: testing suitability for boundary identification in urban areas. GIM Int. 27, 20–25.
72. Hirschmüller, H., 2005. Accurate and efficient stereo processing by semi-global matching and mutual information. IEEE Conf. Comput. Vision Pattern Recognit. 3, 807–814.
73. Hudzietz, B., Saripalli, S., 2011. An experimental evaluation of 3D terrain mapping with an autonomous helicopter. ISPRS – Int. Arch. Photogramm. Remote Sens. Spatial Inform. Sci. XXXVIII-1/C22, 137–142.
74. ICAO, 2011. ICAO Circular 328, Unmanned Aircraft Systems (UAS). Technical Report. International Civil Aviation Authority. Montral, Canada. Imbach, B., Erk, C., 2009. Autonomer UAV Helikopter für Vermessung und industrielle Anwendungen. In: Luhmann, T., Müller, C. (Eds.), Herbert Wichmann, 8. Oldenburger 3D-Tage, pp. 384–389.

75. Israel, M., 2011. A UAV-based roe deer fawn detection system. ISPRS – Int. Arch. Photogramm. Remote Sens. Spatial Inform. Sci. XXXVIII-1/C22, 51–55.
76. Jensen, A.M., Baumann, M., Chen, Y., 2008. Low-cost multispectral aerial imaging using autonomous runway-free small flying wing vehicles. IGARSS 5, 506–509.http://dx.doi.org/10.1109/IGARSS.2008.4780140.
77. Jensen, A.M., Morgan, D., Chen, Y., Clemens, S., Hardy, T., 2009. Using multiple opensource low-cost unmanned aerial vehicles (UAV) for 3D photogrammetry and distributed wind measurement. In: ASME/IEEE 2009 International Conference on Mechatronic and Embedded Systems and Applications; 20th Reliability, Stress Analysis, and Failure Prevention Conference, vol. 3, pp. 629–634. http://dx.doi.org/10.1115/DETC2009-87586.
78. Jensen, A.M., Neilson, B.T., McKee, M., Chen, Y., 2012. Thermal remote sensing with an autonomous unmanned aerial remote sensing platform for surface stream temperatures. IGARSS 5049–5052. http://dx.doi.org/10.1109/IGARSS.2012.6352476
79. Kaess, M., Ranganathan, A., Dellaert, F., 2008. iSAM: incremental smoothing and mapping. IEEE Trans. Robot. 24, 1365–1378.
80. Kalman, R., 1960. A new approach to linear filtering and prediction problems. Trans. ASME J. Basic Eng. 82 (Series D), 35–45.
81. Kelcey, J., Lucieer, A., 2012. Sensor correction and radiometric calibration of a 6- band multispectral imaging sensor for UAV remote sensing. ISPRS – Int. Arch. Photogramm. Remote Sens. Spatial Inform. Sci. XXXIX-B1, 393–398.
82. Kemper, G., 2012. New airborne sensors and platforms for solving specific tasks in remote sensing. ISPRS – Int. Arch. Photogramm. Remote Sens. Spatial Inform. Sci. XXXIX-B5, 351–356.
83. Kendoul, F., 2012. Survey of advances in guidance, navigation, and control of unmanned rotorcraft systems. J. Field Robot. 29, 315–378.
84. Kohoutek, T., Eisenbeiss, H., 2012. Processing of UAV based range imaging data to generate detailed elevation models of complex natural structures. ISPRS – Int. Arch. Photogramm. Remote Sens. Spatial Inform. Sci. XXXIX-B1, 405–410.
85. Kostrzewa, J., Meyer, W., Laband, S., Terre, W., Petrovich, P., Swanson, K., Sundra, C., Sener, W., Wilmott, J., 2003. Infrared microsensor payload for miniature unmanned aerial vehicles. In: Proc. SPIE 5090, Unattended Ground Sensor Technologies and Applications, vol. 265. http://dx.doi.org/10.1117/12.500712.
86. Krzystek, P., 1991. Fully automatic measurement of digital elevation models with MATCH-T. In: 43rd Photogrammetric Week. Institut für Photogrammetrie, Universität Stuttgart, Stuttgart, Germany, pp. 203–214.

87. Küng, O., Strecha, C., Fua, P., Gurdan, D., Achtelik, M., Doth, K.-M., Stumpf, J., 2011. Simplified building models extraction from ultra-light uav imagery. ISPRS – Int. Arch. Photogramm. Remote Sens. Spatial Inform. Sci. XXXVIII-1/C22, 217–222. http://dx.doi.org/10.5194/isprsarchives-XXXVIII-1-C22-217-2011.
88. Leonard, J.J., Durrant-Whyte, H.F., 1991. Simultaneous map building and localization for an autonomous mobile robot, In: International Workshop on Intelligent Robots and Systems '91. Intelligence for Mechanical Systems,Proceedings IROS '91. IEEE/RSJ, 3-5 Nov 1991, vol. 3, pp.1442–1447. http://dx.doi.org/10.1109/IROS.1991.174711.
89. Lowe, D., 2004. Distinctive image features from scale-invariant keypoints. Int. J. Comput. Vision 60, 91–110.
90. Lucieer, A., Robinson, S., Turner, D., Harwin, S., Kelcey, J., 2012. Using a micro-UAV for ultra-high resolution multi-sensor observations of antarctic moss beds. ISPRS – Int. Arch. Photogramm. Remote Sens. Spatial Inform. Sci. XXXIX-B1,429–433.
91. Manyoky, M., Theiler, P., Steudler, D., Eisenbeiss, H., 2011. Unmanned aerial vehicle in cadastral applications. Int. Arch. Photogramm. Remote Sens. Spatial Inf. Sci. XXXVIII-1/C22, 57–62. http://dx.doi.org/10.5194/isprsarchives-XXXVIII-1-C22- 57-2011.
92. MarketsandMarkets, 2013. Unmanned Aerial Vehicle Market (2013–2018). Technical Report. MarketsandMarkets. Dallas, TX, USA.
93. Matsuoka, R., Nagusa, I., Yasuhara, H., Mori, M., Katayama, T., Yachi, N., Hasui, A., Katakuse, M., Atagi, T., 2012. Measurement of large-scale solar power plant by using images acquired by non-metric digital cmaera on board UV. ISPRS – Int. Arch. Photogramm. Remote Sens. Spatial Inform. Sci. XXXIX-B1, 435–440.
94. Mayr, W., 2011a. UAV mapping—a user report. ISPRS – Int. Arch. Photogramm. Remote Sens. Spatial Inform. Sci. XXXVIII-1/C22, 277–282.
95. Mayr, W., 2011b. Unmanned aerial systems in use for mapping at BLOM. In: 53rd Photogrammetric Week. Institut für Photogrammetrie, Universität Stuttgart, pp. 125–134.
96. Mayr, W., 2013. Unmanned aerial systems—for the rest of us. In: 54th Photogrammetric Week. Institut für Photogrammetrie, Universität Stuttgart, pp. 151–163.
97. Meier, L., Tanskanen, P., Fraundorfer, F., Pollefeys, M., 2011. The PIXHAWK pensource
98. computer vision framework for MAVs. ISPRS – Int. Arch. Photogramm. Remote Sens. Spatial Inform. Sci. XXXVIII-1/C22, 13–18.

99. Merz, T., Chapman, S., 2011. Autonomous unmanned helicopter system for remote sensing missions in unknown environments. ISPRS – Int. Arch. Photogramm. Remote Sens. Spatial Inform. Sci. XXXVIII-1/C22, 143–148.
100.Mészáros, J., 2011. Aerial surveying UAV based on open-source hardware and software. ISPRS – Int. Arch. Photogramm. Remote Sens. Spatial Inform. Sci. XXXVIII-1/C22.
101.Miraliakbari, A., Hahn, M., Engels, J., 2012. Vibrations of a gyrocopter—an analysis using IMUs. ISPRS – Int. Arch. Photogramm. Remote Sens. Spatial Inform. Sci. XXXIX-B1, 497–502.
102.Molina, P., Parés, M., Colomina, I., Vitoria, T., Silva, P., Skaloud, J., Kornus, W., Prades, R., Aguilera, C., 2012. Drones to the Rescue! unmanned aerial search missions based on thermal imaging and reliable navigation. InsideGNSS 7, 36–47.
103.Morel, J., Yu, G., 2009. ASIFT: A new framework for fully affine invariant image comparison. SIAM J. Imaging Sci. 2, 438–469.
104.Nagai, M., Shibasaki, R., Manandhar, D., Zhao, H., 2004. Development of Digital Surface and Feature Extraction by Integrating Laser Scanner and CCD Sensor with IMU. ISPRS – Int. Arch. Photogramm. Remote Sens. Spatial Inform. Sci. XXXV-B5, 655–659.
105.Advanced Navigation, 2012. Spatial Performance Evaluation. Technical Report. Advanced Navigation. Sydney, NSW, Australia.
106.Niethammer, U., Rothmund, S., Schwaderer, U., Zeman, J., Joswig, M., 2011. Open source image-processing tools for low-cost UAV-based landslide investigations. ISPRS – Int. Arch. Photogramm. Remote Sens. Spatial Inform. Sci. XXXVIII-1/C22,57–62.
107.Novatel, 2013. SPAN OEM-IMU-ADIS-16488: Commercial MEMS IMU Integrated with SPAN Technology to Deliver 3D Position, Velocity and Attitude. Novatel Inc. Calgary, AB, Canada.
108.van Persie, M., Oostdijk, A., Fix, J., van Sijl, M., Edgardh, L., 2011. Real-time UAV based geospatial video integrated into the fire brigades crisis management GIS system. ISPRS – Int. Arch. Photogramm. Remote Sens. Spatial Inform. Sci. XXXVIII-1/C22, 173–175.
109.Petrie, G., 2013. Commercial operation of lightweight UAVs for aerial imaging and mapping. GEOInformatics 16, 28–39.
110.Pierrot Deseilligny, M., Clery, I., 2011. APERO, an open source bundle adjustment software for automatic calibration and orientation of set of images. Int. Arch. Photogramm. Remote Sens. Spatial Inf. Sci. XXXVIII-5/W16, 269–276. http://dx.doi.org/10.5194/isprsarchives-XXXVIII-5-W16-269-2011.
111.Przybilla, H., Wester-Ebbinghaus, W., 1979. Bildflug mit ferngelenktem Kleinflugzeug. Bildmessung und Luftbildwessen 47, 137–142.

112.Qin, R., Grün, A., Huang, X., 2013. UAV project – building a reality-based 3D model. Coordinates 9, 18–26.

113.Rau, J., Jhan, J., Lob, C., Linb, Y., 2011. Landslide mapping using imagery acquired by a fixed-wing UAV. ISPRS – Int. Arch. Photogramm. Remote Sens. Spatial Inform. Sci. XXXVIII-1/C22, 195–200.

114.Rehak, M., Mabillard, R., Skaloud, J., 2013. A micro-UAV with the capability of direct georeferencing. ISPRS – Int. Arch. Photogramm. Remote Sens. Spatial Inform. Sci. XL-1/W2, 317–323.

115.Remondino, F., 2011. Heritage recording and 3D modeling with photogrammetry and 3D scanning. Remote Sens. 3, 1104–1138.

116.Remondino, F., Barazzetti, L., Nex, F., Scaioni, M., Sarazzi, D., 2011. UAV photogrammetry for mapping and 3D modelling—current status and future perspectives. ISPRS – Int. Arch. Photogramm. Remote Sens. Spatial Inform. Sci., 25–31.

117.Remondino, F., del Pizzo, S., Kersten, T., Troisi, S., 2012. Low-cost and open-source solutions for automated image orientation – a critical overview. In: Ioannides, M., Fritsch, D., Leissner, J., Davies, R., Remondino, F., Caffo, R. (Eds.), Progress in Cultural Heritage Preservation, Lecture Notes in Computer Science, vol. 7616.Springer, Berlin Heidelberg, pp. 40–54.

118.Remy, M., de Macedo, K., Moreira, J., 2012. The First UAV-based P- and X-band Interferometric SAR System. IEEE, München, Germany, pp. 5041–5044.

119.Rikola, Ltd., 2012. Rikola Hyper-spectral camera specifications, accessed on February 2014.

120.Rinaudo, F., Chiabrando, F., Lingua, A., Span, A., 2012. Archaeological site monitoring: UAV photogrammetry can be an answer. ISPRS – Int. Arch. Photogramm. Remote Sens. Spatial Inform. Sci. XXXIX-B5, 583–588.

121.Rosen, P.A., Hensley, S., Wheeler, K., Sadowy, G., Miller, T., Shaffer, S., Muellerschoen, R., Jones, C., Zebker, H., Madsen, S., 2006. UAVSAR: a new NASA airborne SAR system for science and technology research, In: 2006 IEEE Conference on Radar, 24–27 April 2006, p. 8. http://dx.doi.org/10.1109/RADAR.2006.1631770.

122.Rosnell, T., Honkavaara, E., 2012. Point cloud generation from aerial image data acquired by a quadrocopter type micro unmanned aerial vehicle and a digital still camera. Sensors 12, 453–480.

123.Rufino, G., Moccia, A., 2005. Integrated VIS-NIR Hyperspectral/thermal-IR lectrooptical Payload System for a Mini-UAV. American Institute of Aeronautics and

124.Astronautics, Arlington, VA, USA, pp. 647–664.

125.Scherer, S., Singh, S., Chamberlain, L., Elgersma, M., 2008. Flying fast and low among obstacles: methodology and experiments. Int. J. Robot. Res. 27, 549–574.

126.Schinstock, D., Lewis, C., Buckley, C., 2009. An alternative cost function to bundle adjustment used for aerial photography from UAVs. In: Proceedings of the ASPRS Annual Conference 2009, American Society for Photogrammetry and Remote Sensing (ASPRS), Baltimore, Maryland, March 9–13, 2009.

127.Scholtz, A., Kaschwich, C., Kruger, A., Kufieta, K., Schnetter, P., Wilkens, C., Kruger, T., Vorsmann, P., 2011. Development of a new multi-purpose UAS for scientific application. ISPRS – Int. Arch. Photogramm. Remote Sens. Spatial Inform. Sci. XXXVIII-1/C22, 149–154.

128.Schulz, H., 2011. The unmanned mission avionics test helicopter—a flexible and versatile VTOL-UAS experimental system. ISPRS – Int. Arch. Photogramm. Remote Sens. Spatial Inform. Sci. XXXVIII-1/C22, 309–314.

129.Seitz, C., Altenbach, H., 2011. Project ArchEye – the quadracopter as the archaelogist's eye. ISPRS – Int. Arch. Photogramm. Remote Sens. Spatial Inform. Sci. XXXVIII-1/C22, 297–302.

130.Shi, J., Wang, J., Xuc, Y., 2011. Object-based change detection using georeferenced UAV images. ISPRS – Int. Arch. Photogramm. Remote Sens. Spatial Inform. Sci. XXXVIII-1/C22, 177–182.

131.Singh, R., Jain, A., Sen, P., Radhakrishnan, V., Kumar, K., 2012. Autonomous unmanned aerial vehicle for reconnaissance based on robotic operating system. In: Third Symposium on Indoor Flight Issues. Birla Institute of Technology and Science, Dubai, UAE, p. 15.

132.Smith, R., Self, M., Cheeseman, P., 1986. Estimating uncertain spatial relationships in robotics. In: Proceedings of the Second Conference Annual Conference on Uncertainty in Artificial Intelligence (UAI-86). AUAI Press, Corvallis, OR, USA, pp. 267–288.

133.Snavely, N., Seitz, S.M., Szeliski, R., 2008. Modeling the world from internet photo collections. Int. J. Comput. Vision 80, 189–210.

134.Stødle, P., Borch, N., Storvold, R., 2013. High-performance Visualization of UAS Sensor and Image Data with Raster Maps and Topography in 3D. ISPRS - Int. Arch. Photogram. Rem. Sens. Spat. Inform. Sci. XL-1/W2, 275–280. http://dx.doi.org/10.5194/isprsarchives-XL-1-W2-275-2013.

135.Strasdat, H., Montiel, J., Davison, A., 2012. Visual SLAM: Why filter? Image Vision Comput. 30, 65–77.

136.Strecha, C., 2011. Automated photogrammetric techniques on ultra-light UAV imagery. In: 53rd Photogrammetric Week. Institut für Photogrammetrie, Universität Stuttgart, pp. 289–294.

137. Strecha, C., Bronstein, A., Bronstein, M., Fua, P., 2012. LDAHash: Improved matching with smaller descriptors. IEEE Trans. Pattern Anal. Mach. Intell. 34, 66–78.
138. Tetracam, 2012. MiniMCA.
139. Thamm, H., 2011. Susi 62: A robust and safe parachute UAV with long flight time and good payload. ISPRS – Int. Arch. Photogramm. Remote Sens. Spatial Inform. Sci. XXXVIII-1/C22, 19–24.
140. Tsingas, V., 1992. Automatisierung der Punktübertragung in der Aerotriangulation durch mehrfache digitale Zuordnung. Ph.D. thesis. Universität Stuttgart. Germany. (Published by the Deutsche Geodätische Kommission, Reihe C, No.392, München, Germany).
141. UAVMarketSpace, 2012. Autopilot Navigation.
142. Vallet, J., Panissod, F., Strecha, C., Tracol, M., 2011. Photogrammetric performance of
143. an ultra light weight swinglet UAV. ISPRS – Int. Arch. Photogramm. Remote Sens. Spatial Inform. Sci. XXXVIII-1/C22, 253–258. http://dx.doi.org/10.5194/ isprsarchives-XXXVIII-1-C22-253-2011.
144. van Blyenburgh, P., 2013. 2013–2014 RPAS Yearbook: Remotely Piloted Aircraft Systems: The Global Perspective 2013/2014. Technical Report. UVS International. Paris, France.
145. Wallace, L., Lucieer, A., Watson, C., Turner, C., 2012. Assessing the feasibility of UAVbased LiDAR for high resolution forest change detection. ISPRS – Ann. Photogramm. Remote Sens. Spatial Inform. Sci. XXXIX-B7, 499–504.
146. Wang, J., Garratt, M., Lambert, A., Wang, J.J., Han, S., Sinclair, D., 2008. Integration of GPS/INS/vision sensors to navigate unmanned aerial vehicles. ISPRS – Int. Arch. Photogramm. Remote Sens. Spatial Inform. Sci. XXXVII-B1, 963–970.
147. Wefelscheid, C., Hansch, R., Hellwich, O., 2011. Three-dimensional building reconstruction using images obtained by unmanned aerial vehicles. ISPRS – Int. Arch. Photogramm. Remote Sens. Spatial Inform. Sci. XXXVIII-1/C22, 183–188.
148. Wester-Ebbinghaus, W., 1980. Aerial photography by radio controlled model helicopter. Photogramm. Rec. 10, 85–92.
149. Wich, S., Koh, L., 2012. Conservation drones: the use of unmanned aerial vehicles by ecologists. GIM Int. 26, 29–33.
150. Xie, F., Lin, Z., Gui, D., Lin, H., 2012. Study on construction of 3D building based on UAV images. ISPRS – Int. Arch. Photogramm. Remote Sens. Spatial Inform. Sci. XXXIX-B1, 469–473.

151.Yun, M., Kimb, J., Seo, D., Lee, J., Choi, C., 2012. Application possibility of smartphone as payload for photogrammetric UAV system. ISPRS – Int. Arch. Photogramm. Remote Sens. Spatial Inform. Sci. XXXIX-B4, 349–352.

152.Zarco-Tejada, P., Berni, J., 2012. Vegetation monitoring using a micro-hyperspectral imaging sensor onboard an unmanned aerial vehicle (UAV). In: Proceedings of the EuroCOW 2012, European Spatial Data Research (EuroSDR), Castelldefels, Spain.

153.Zhou, G., Yang, J., Li, X., Yang, X., 2012. Advances of flash LiDAR development onboard UAV. ISPRS – Ann. Photogramm. Remote Sens. Spatial Inform. Sci. XXXIX-B3, 193–198.

CITATION

I. Colomina, P. Molina, Unmanned aerial systems for photogrammetry and remote sensing: A review, ISPRS Journal of Photogrammetry and Remote Sensing, Volume 92, June 2014, Pages 79-97, ISSN 0924-2716, http://dx.doi.org/10.1016/j.isprsjprs.2014.02.013.

CHAPTER 2

A Review of Ocean Color Remote Sensing Methods and Statistical Techniques for the Detection, Mapping and Analysis of Phytoplankton Blooms in Coastal and Open Oceans

David Blondeau-Patissier[1], James F.R. Gower[2], Arnold G. Dekker[3,4] , Stuart R. Phinn[4], Vittorio E. Brando[3, 4]

[1] North Australian Marine Research Alliance (NAMRA), The Research Institute for the Environment and Livelihoods (RIEL), Charles Darwin University, Darwin, Australia

[2] Fisheries and Oceans Canada, Institute of Ocean Sciences, Sidney, Canada

[3] CSIRO Land and Water, Aquatic Remote Sensing Group, Canberra, Australia

[4] School of Geography, Planning and Environmental Management, The University of Queensland, Brisbane, Australia

ABSTRACT

The need for more effective environmental monitoring of the open and coastal ocean has recently led to notable advances in satellite ocean color technology and algorithm research. Satellite ocean color sensors' data are widely used for the detection, mapping and monitoring of phytoplankton blooms because earth observation provides a synoptic view of the ocean, both spatially and temporally. Algal blooms are indicators of marine ecosystem health; thus, their monitoring is a key component of effective management of coastal and oceanic resources. Since the late 1970s, a wide variety of operational ocean color satellite sensors and algorithms have been developed. The comprehensive review presented in this article captures the details of the progress and discusses the advantages and limitations of the algorithms used with the multi-spectral ocean color sensors CZCS, SeaWiFS, MODIS and MERIS. Present challenges include overcoming the severe limitation of these algorithms in coastal waters and refining detection limits in various oceanic and coastal environments. To understand the spatio-temporal patterns of algal blooms and their triggering factors, it is essential to consider the possible effects of environmental parameters, such as water temperature, turbidity, solar radiation and bathymetry. Hence, this review will also discuss the use of statistical techniques and additional datasets derived from ecosystem models or other satellite sensors to characterize further the factors triggering or limiting the development of algal blooms in coastal and open ocean waters.

INTRODUCTION

Over 5000 species of marine phytoplankton have been described worldwide (e.g.,Sournia et al., 1991). Typically ranging from less than 1 μm to over 100 μm in size, a phytoplankton cell, also known as an 'algal' or 'algae' cell, is a planktonic photosynthesizing organism. Increases in phytoplankton cell numbers can result from favorable environmental conditions, which include water column stratification, increase in light availability (e.g., Gohin et al., 2003 and Kogeler and Rey, 1999), water temperature (Thomas et al., 2003) and/or nutrient levels (e.g., Santoleri et al., 2003 and Siegel et al., 1999). The global distribution of Chlorophyll-a (Chl-a), the direct proxy for phytoplankton biomass (Cullen, 1982), shows that Chl-a–rich regions are located along the coasts and continental shelves, north of 45° North (Fig. 1a), mostly because of a strong nutrient supply. Moderate Chl-a concentrations are found in the equatorial regions of the Atlantic and Pacific, caused by the upwelling of deep, nutrient-rich, cool waters from the divergence of the ocean water masses along the equator. Moderate Chl-a concentrations are also found in the subtropical convergence zone (south of 45° South), where cool, nutrient-rich sub-Antarctic water masses mix with warm, nutrient-poor subtropical waters. However, most open ocean regions typically appear low in satellite-derived Chl because they are far from land. Ocean color observations are limited to the first optical depth; consequently, deep chlorophyll maxima (DCM) are not always captured by satellites (e.g., Huisman et al., 2006 and Cullen, 1982). Many phytoplankton blooms (see Section 2) occurring deep in the water column or with extremely low Chl-a (<0.1 mg m^{-3}) remain unreported because they are not always observed in satellite images but yet are known to occur (e.g., Dore et al.,

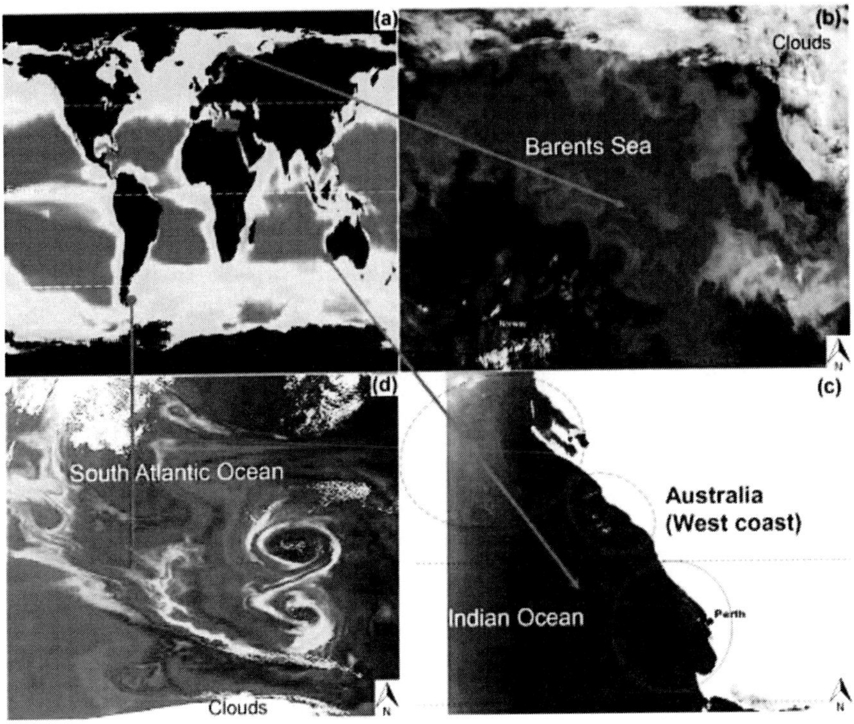

Figure 1. (a) The global distribution of chlorophyll averaged over the period from 1 January 2002 to 31 January 2008 using data collected from MODIS-Aqua; Chlorophyll values range from 0.01 mg m^{-3} (purple) to 60 mg m^{-3}(red) (NASA/GSFC); (b) NASAMODIS-Aqua captures a Coccolithophore bloom in the Barents Sea on August 14, 2011 (credit: NASA); (c) ESAMERIS reduced-resolution RGB-stretched image showing surface bloom expressions (circled in red) off the Australian West coast on March 10, 2009 (fromBlondeau-Patissier, 2011); (d) ESAMERIS full-resolution (300 m) true-color image showing a southern hemisphere spring phytoplankton bloom (green swirls) in the South Atlantic Ocean, 600 km east off the Falkland Islands, on December 2, 2011. "The figure-eight pattern in the image is likely to be an example of von Karman vortices, which are formed when an obstacle blocks a prevailing wind or ocean current" (credit: ESA). (For interpretation of the references to colour in this figure legend, the reader is referred to the web version of this article.)

2008 and Villareal et al., 2011). Algal blooms (see Section 2) detected by satellite sensors often cover large areas, but their typically "patchy" distributions make them difficult to model (Martin, 2003) (Fig. 1b–d). The visualization of satellite images (Fig. 1) is the primary technique used to identify their presence, particularly when phytoplankton blooms occur as a regular event in a specific ocean region (e.g., Srokosz and Quartly, 2013) or in regions where they are not usually expected, such as oligotrophic gyres (e.g.,Wilson, 2003, Wilson et al., 2008 and Wilson and Qiu, 2008). Over the last 13 years, there has been an increase in peer-reviewed publications on the study of algal blooms using ocean color satellite data (Fig. 2). Algal blooms in coastal ocean regions have been the primary focus of those studies, mainly due to the direct connectivity between the land and continental shelf waters (e.g., Gazeau et al., 2004) and the impact of coastal harmful phytoplankton blooms on anthropogenic activities (Frolov et al., 2013). Important technological progress in the design of satellite ocean color sensors from the second and third generations greatly improved coastal water algorithms, resulting in a more accurate retrieval of phytoplankton proxies in coastal waters (see Table 1 of Shen et al. (2012);Fig. 3).

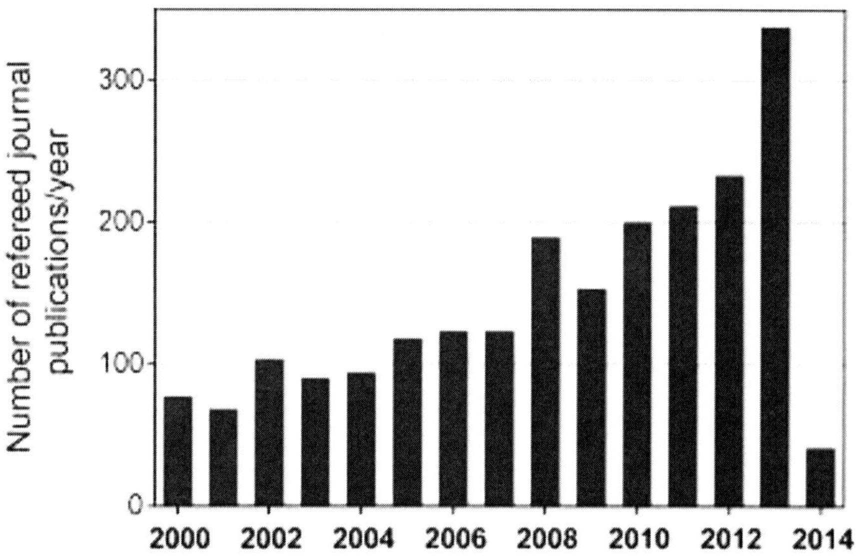

Figure 2. Number of peer-reviewed journal publications per year since 2000 based on the results of a Boolean search with the terms "ocean color*" and "phytoplankton algal bloom*" in ScienceDirect® (as of December16, 2013). Data for the year 2014 refer to accepted publications that are available online and expected to be published between January and February 2014.

Phytoplankton blooms affect the color of the water by increasing light backscattering with spectrally localized water-leaving radiance minima from generic (Chl-a) and species-specific algal pigment absorption, such as phycobiliproteins for cyanobacteria, fucoxanthin for diatoms and peridinin for dinoflagellates. Ocean color remote sensing, the passive satellite-based

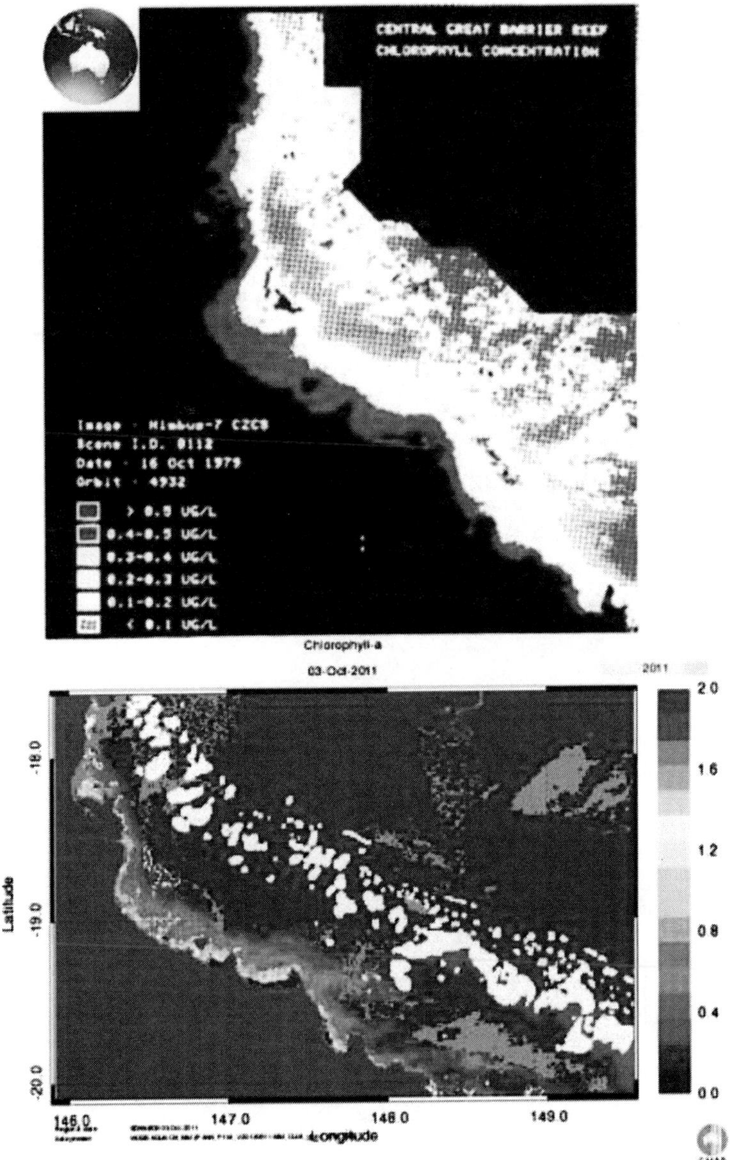

Figure 3. Examples of Chl-a satellite images from the first and second generations of ocean color sensors. Top panel: a CZCS image on October 16, 1979 from Cairns to Broad Sound, Great Barrier Reef, Australia (Gabric et al., 1990); Bottom panel: the same region as observed from MODIS-Aqua on October 3, 2011. Chl-a was derived using the Matrix Inversion Method (Brando et al., 2012) and SeaDAS 6.4. Clouds and Land are masked in gray and the submerged reef matrix is masked in white. Chl-a is expressed in μg/L. (For interpretation of the references to colour in this figure legend, the reader is referred to the web version of this article.)

measurement of visible light emerging from the ocean surface (Robinson, 2004), has provided more than two decades of near-real-time synoptic and recurrent measurements of global phytoplankton biomass (Fig. 1 and Fig. 4), evolving from qualitative (e.g., Gordon et al., 1980) to quantitative estimates (e.g.,Kutser, 2009). The accumulation of scientific knowledge on the temporal and spatial dynamics of phytoplankton in the world's oceans from earth observations was largely assisted by rapid advances in marine science technologies (e.g., Babin et al., 2008 and Dickey et al., 2006) and has had many global and local applications (Table 1). Recent reviews on satellite ocean color remote sensing have reported on, but are not limited to, scientific advances in this field (e.g., McClain, 2009a and McClain, 2009b), its societal benefits (e.g., IOCCG Report 7, 2008) and its valuable applications to coastal ecosystem management (e.g., Kratzer et al., 2013 and Klemas, 2011), including that of fisheries (e.g., Wilson, 2011) and in the detection of (harmful) algal blooms (Shen et al., 2012) (Table 2).

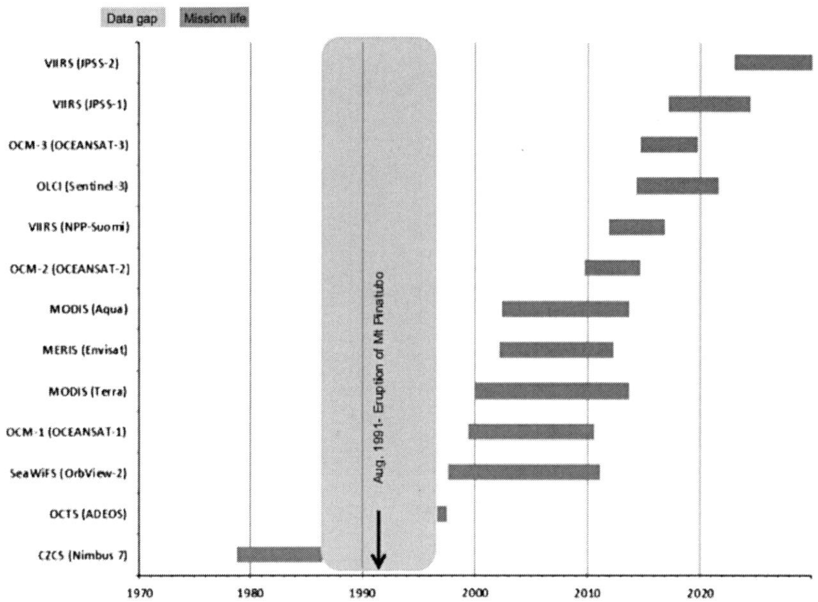

Figure 4. Timeline 1970–2030 illustrating past, current, and future global ocean-color satellite missions. Missions after 1999 were extracted from the online CEOS Earth Observation Handbook (http://www.eohandbook.com/). Satellite platforms are indicated. (For interpretation of the references to colour in this figure legend, the reader is referred to the web version of this article.)

Table 1. Detecting algal blooms from ocean color remote sensing: applications overview (cited research from 2006 to 2013 is indicative only).

Research application	Global scale	Local scale
Spatial and temporal distributions (phenology)	Gower and King, 2011a,Gower et al., 2008,Demarcq et al., 2012 and Racault et al., 2012	Park et al., 2010b, Garcia and Garcia, 2008, Gower and King, 2007a, Song et al., 2010, Platt and Sathyendranath, 2008 and Henson et al., 2009
Derivation of long terms baseline, and marine ecosystem's response to i.e., climatic, anthropogenic forcing	Martinez et al., 2009 and Siegel et al., 2013	Kahru et al., 2010, Shi and Wang, 2007 and Zhao et al., 2008
Ecosystem partitioning	IOCCG, 2009, Oliver and Irwin, 2008 and Platt et al., 2008	Henson et al. (2006)
Coastal zone management (eutrophication, etc.)	Smetacek and Cloern (2008)	Peñaflor et al., 2007, Banks et al., 2012 and Klemas, 2011
Biogeochemistry (carbon, nitrogen fluxes, etc.)	Platt et al. (2008)	Chang and Xuan, 2011 and Focardi et al., 2009
Major phytoplankton groups (Coccolithophores,*Tric hodesmium*, etc.)	Moore et al. (2012)	Carvalho et al., 2011, Miller et al., 2006,Shutler et al., 2012, McKinna et al., 2011 and Gower and King, 2011b
Size-class, community composition and physiology	Brewin et al., 2011 and Behrenfel d et al., 2009 and	Pan et al., 2012 and Pan et al., 2013)
Extremes (e.g., super-blooms)	Gower and King (2011a)	Gower and King (2007a)

Table 2. Spectral bands (in nm) used in algal bloom indices for SeaWiFS, MODIS and MERIS (cited research is indicative only).

Sensor	Product	Band 1	Band 2	Band 3	Algal bloom type	References
MODIS	FLH	667	678	746	Algal blooms, surface algal blooms	Hu et al. (2005)
MERIS	FLH	665	681	708.75	Algal blooms	Gower et al. (2003)
MODIS	FAI	667	859	1240 or 1640	Surface algal blooms	Hu (2009)
SeaWiFS	CIA	443	555	670	Low Chlorophyll concentrations	Hu et al. (2012)
MERIS	MCI	681	708.75	753	High concentration in water and surface algal blooms	Gower et al. (2003)
MERIS	MCI_{wide}	665	708.75	753	High concentration in water and surface algal blooms	Gower et al. (2004)
MERIS	EBI	665	708.75	n/a	Surface algal blooms	(Not peer-reviewed) Amin, R.
SeaWiFS	RI	443	510	555	Algal blooms	Ahn and Shanmugam (2006)
SeaWiFS	ABI	443	490	555	HAB, algal blooms	Shanmugam, 2011 and Ahn and Shanmugam, 2006
MODIS	ABI	443	490	555	HAB, algal blooms	Shanmugam, 2011 and Ahn and Shanmugam, 2006
MODIS	KBBI & RBD	667	678	n/a	Surface algal blooms (*K. brevis*)	Amin et al. (2009b)
MERIS	KBBI & RBD	665	681	n/a	Surface algal blooms (*K. brevis*)	Amin et al. (2008)

The reviews of Matthews (2011) and Odermatt et al. (2012) discussed the various ocean color models used for the retrieval of water quality parameters in open and coastal ocean waters, from empirical to more complex approaches. More recently, Brody et al. (2013)compared different methods to determine phytoplankton bloom initiation. The present article will complement those recent reviews by providing the following:

1. A comparative analysis of algorithm types specifically used for the detection, mapping and analysis of algal blooms from passive multi-spectral ocean color sensors' imagery. Advantages and limitations in the application of the techniques presented will be discussed.
2. A synthesis of the ocean color remote sensing methods used for the detection of four specific types of blooms. An exception is made for pelagic *Sargassum*, which are macro-algae and not typically defined as phytoplankton.

3. An examination of the statistical techniques that are often used in combination with an ocean color dataset for the detection and monitoring of algal blooms.

The detection of phytoplankton functional types, phytoplankton size classes and phytoplankton primary production derived from ocean color remote sensing data will not be reviewed in this article. Table 1, Table 2, Table 3, Table 4, Table 5 and Table 6 classify the peer-reviewed publications published between 2000 and 2014 according to the research application, the ocean color sensor used, the phytoplankton bloom types and the techniques employed to detect those events.

Table 3. Detecting phytoplankton blooms with CZCS (cited research is indicative only).

Phytoplankton type	Reflectance classification (thresholds, anomalies)	Reflectance band-ratios	Bio-optical model or neural network	Spectral band difference	Satellite product (threshold or anomaly) and climatology, statistics
Phytoplankton bloom (undefined)		Banse and English, 2000, Kim et al., 2000,Yoder et al., 2001,Shevyrnogov et al., 2002b, Robinson et al., 2004 and Tang et al., 2004			Banse and English, 1999,Shevyrnogov et al., 2002a,Marinelli et al., 2008 and Antoine et al., 2005[a]
Surface expression (undefined)					
Coccolithophores	Merico et al. (2003)				
Trichodesmium					
Red tides/HAB					

ALGAL BLOOMS IN THE CONTEXT OF THIS REVIEW

The various definitions of phytoplankton blooms often rely on different, and sometimes arbitrary, criteria, such as the biomass or growth rate, or both. The review article byKutser (2009) warned in the introduction that the term "bloom" is "*relative*" because it is used to describe phytoplankton events with contrastingly different biomass concentrations (p. 4402). Richardson (1997) defined an algal bloom as "*the rapid growth of one or more species which leads to an increase in biomass of the species*" (p. 302). In the context of this review, a phytoplankton bloom will be

Table 4. Detecting phytoplankton blooms with SeaWiFS (cited research is indicative only).

Phytoplankton type	Reflectance classification (thresholds, anomalies)	Reflectance band-ratios	Bio-optical model or neural network	Spectral band difference	Satellite product (threshold or anomaly) and climatology, statistics
Phytoplankton bloom (undefined)	Otero and Siegel (2004)	Gower (2001), Lavender and Groom (2001), Gohin et al. (2003), Santoleri et al. (2003), Thomas et al. (2003), Vinayachandran and Mathew (2003), Claustre and Maritorena (2003), Babin et al. (2004), Otero and Siegel (2004), Ahn et al. (2005), Iida and Saitoh (2007) and Siegel et al. (2013)	Shanmugam (2011), Dierssen and Smith (2000) and Siegel et al. (2013)	Hu et al. (2012)	Saitoh et al. (2002), Nezlin and Li (2003), Strokosz et al. (2004), Brickley and Thomas (2004), Navarro and Ruiz (2006), Tan et al. (2006), Henson and Thomas (2007), Marrari et al. (2008), Yoo et al. (2008), Garcia-Soto and Pingree (2009), Henson et al. (2009), Platt et al. (2009) Quartly and Srokosz (2004), Vargas et al. (2009), Song et al. (2010), Raitsos et al. (2011), Racault et al. (2012) and Kilston et al. (2013)
Surface expression (undefined)					
Coccolithophores	Iida et al. (2002), Zeichen and Robinson (2004) and Moore et al. (2012)				
Trichodesmium	Subramaniam et al. (2002) and Dupouy et al. (2011)				
Red tides/HAB		Ahn et al. (2006) and Tang et al. (2006)	Stumpf (2001)	Ahn and Shanmugam (2006)and Shanmugam et al. (2008)	Miller et al. (2006), Stumpf et al. (2003) and Tomlinson et al. (2004)

Table 5. Detecting phytoplankton blooms with MODIS (cited research is indicative only).

Phytoplankton type	Reflectance classification (thresholds, anomalies)	Reflectance band-ratios	Bio-optical model or neural network	Spectral band difference	Satellite product (threshold or anomaly) and climatology, statistics
Phytoplankton bloom (undefined)		Venables et al. (2007)[a],	Shang et al. (2010), Shanmugam (2011) and Mélin et al. (2011)[a]	Hu et al. (2012)	Wang and Zhao (2008)[a], Shi and Wang (2007), Uz (2007)[a], Peñaflor et al. (2007), White et al. (2007)[a], Oliveira et al. (2009), Park et al. (2010a, 2010b), Raj et al. (2010)[a], Wang et al. (2010)[a], Li et al. (2010b) and
		Acker et al. (2008)[a], Zhao et al. (2009a)[a], Kahru et al. (2010)[a], Nezlin et al. (2010)[a] and Le et al. (2013)			Dasgupta et al. (2009)
Surface expression (undefined)					
Coccolithophores	Signorini and McClain (2009), Iida et al. (2012) and Moore et al. (2012)	Balch et al. (2005)			
Trichodesmium	McKinna et al. (2011)			Hu et al. (2010a)	
Red tides/HAB	Siswanto et al (2013)	Kahru et al. (2004) and Carvalho et al. (2011)	Cannizzaro et al. (2008)[a] and Carvalho et al. (2010)	Hu et al. (2005), Ryan et al. (2009) and Zhao et al. (2010)[a]	Cannizzaro et al. (2008), Cannizzaro (2004), Hu et al. (2004), Tomlinson et al. (2008)[a], Anderson et al. (2011), Banks et al. (2012)[a], Shutler et al (2012) and Kurekin et al. (2014)[a]

[a]Merged with SeaWiFS.

[b]Merged with MERIS.

Table 6. Detecting phytoplankton blooms with MERIS (cited research is indicative only).

Phytoplankton type	Reflectance classification (thresholds, anomalies)	Reflectance band-ratios	Bio-optical model or neural network	Spectral band difference	Satellite product (threshold or anomaly) and climatology, statistics
Phytoplankton bloom (undefined)			Uiboupin et al. (2012)	Gower et al., 2008 and Gower et al., 2005	Gordoa et al. (2008)
Surface expression (undefined)				Gower and King (2011a)	
Coccolithophores	Moore et al. (2012)				
Trichodesmium				Matthews et al. (2012)	
Red tides/HAB		Bernard et al. (2005)		Ryan et al. (2008) andJessup et al. (2009)	Li et al. (2010a)

defined as a biological event composed of micro-algal species (see Bricaud et al. (2004) for phytoplankton cell sizes and ranges of Chl-a concentrations) that is sustained both over time and space and that results in noticeable changes in satellite radiances at wavelengths used for algal bloom proxies due to an increase in biomass (in comparison to surrounding algal bloom-free waters).

OCEAN COLOR REMOTE SENSING ALGORITHMS

The need for accurate retrievals of Chl concentrations in open and coastal ocean waters from ocean color data has driven most of the research in algorithm development over the past thirty years. Other algorithms have also been developed, such as those that use specific spectral features of the reflectance spectrum to detect phytoplankton with surface expressions. There is a wide variety of operational ocean color satellite sensors and algorithms to assist in the detection and monitoring of phytoplankton blooms, and this section explores the various forms currently available, which specifically includes reflectance-based classification algorithms, spectral band-ratios, spectral band-difference algorithms and bio-optical models. The limitations and advantages associated with their application in the detection and mapping of algal blooms are discussed.

Reflectance classification algorithms

It has long been recognized that information about optically active constituents present within a parcel of water can be obtained from its spectral reflectance spectrum (e.g.,Steemann Nielsen, 1963, Steemann Nielsen, 1937 and Steemann Nielsen and Jensen, 1957). Spectral bands located in the blue, green, yellow, red or near-infrared (NIR) portion of the reflectance spectrum can be used in many ways to detect algal blooms (Fig. 5 and Fig. 6). Algorithms that rely mostly on the detection of specific spectral features are often well suited for algal blooms with surface expressions. This approach can be sufficient for the discrimination of algal blooms from other naturally occurring phenomena (e.g., Siegel et al., 2007), but the sole use of the reflectance spectrum can often only provide qualitative estimates. The reliability of the measured reflectance is hampered by its sensitivity to, e.g., the thickness of the floating algal layer, suspended particulates, bottom reflectance (although novel correction techniques now exist (e.g.,Barnes et al., 2013)) and atmospheric correction errors. This reliability is even more questionable when dealing with coastal waters, where other optically active substances affect the water-leaving radiance. To ensure the validity of the algal bloom information retrieved from reflectance classification algorithms, it is recommended that knowledge of the study region be taken into account. A detailed analysis of the reflectance spectra of the flagged pixels is also required. Section 4.1 provides further discussion for the specific detection of Coccolithophore and Trichodesmium blooms.

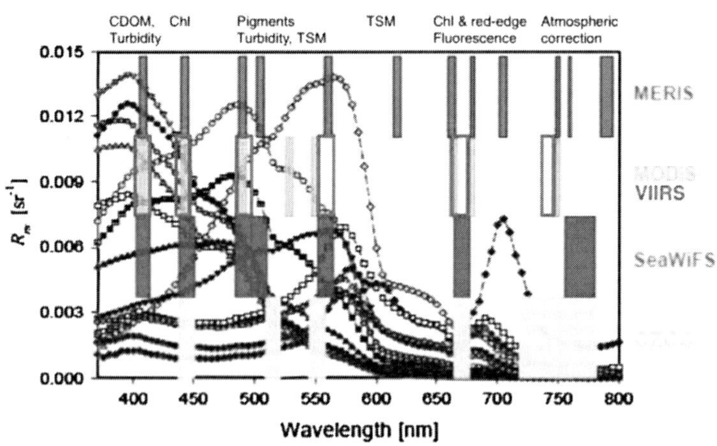

Figure 5. Comparison of the spectral band positions for five ocean color sensors of the first (CZCS), second (SeaWiFS), third (MERIS, MODIS) and fourth (VIIRS) generations. (Figure modified from Fig. 1 of Lee et al. (2007). The authors used 14 remote sensing

reflectance spectra from various waters around the world. The potential applications for each spectral region are indicated. See Table 2 as well (this review.) (For interpretation of the references to colour in this figure legend, the reader is referred to the web version of this article.)

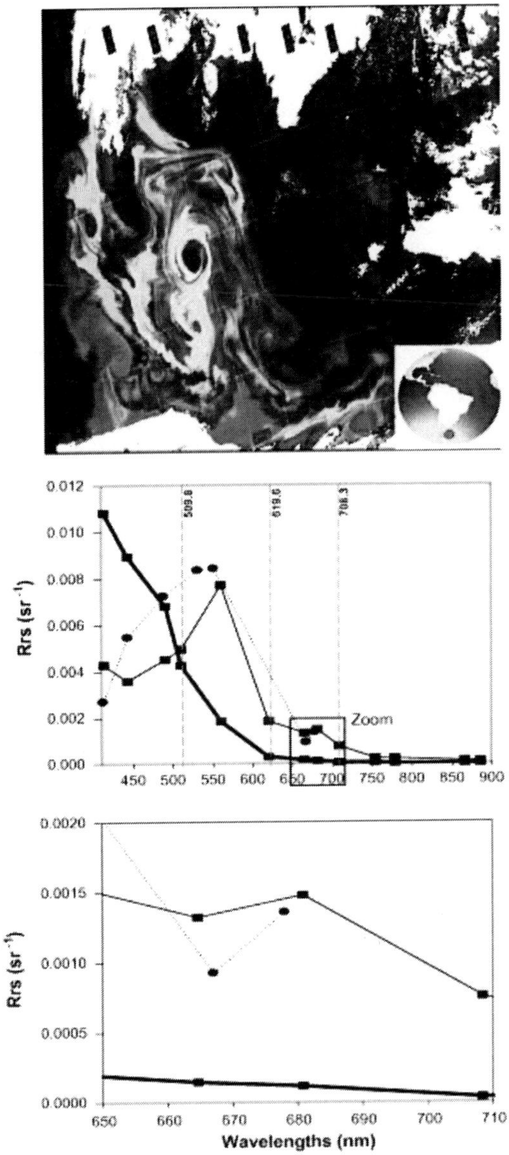

Figure 6. Detecting algal blooms from MODIS and MERIS ocean color sensors. Top panel: RGB MERIS image of December 2, 2011 showing the pixel location (red dot) for the extracted reflectance data; middle panel: remote sensing reflectance spectra (Rrs) of MERIS (thin line) and MODIS (dotted line) Level 2

(standard algorithms) for the same day. Reflectance from the surrounding blue water is shown (thick line); bottom panel: zoom on the spectral region 650–710 nm. MERIS bands at 509.81, 619.30 and 708.32 nm are shown by vertical dashed lines. (For interpretation of the references to colour in this figure legend, the reader is referred to the web version of this article.)

Reflectance band-ratio algorithms

In open ocean Case 1 waters, phytoplankton is the primary water constituent (Morel, 1980 and Morel and Prieur, 1977); thus, Chl-a concentrations can be empirically related to the water-leaving reflectance using relationships of various forms (e.g., Matthews, 2011 and Dierssen, 2010). These empirical relationships are often derived using large, sometimes global (e.g., Fargion and McClain, 2003), *in situ* datasets of coincident Chl-a and reflectance measurements. Empirical blue–green (440–550 nm) spectral band-ratios are the most common types of ocean color algorithms used for Chl-a retrievals because most of the phytoplankton absorption occurs within this portion of the visible spectrum. However, the use of visible wavelengths can be unreliable in coastal waters. In optically complex, Case 2 waters, blue–green reflectance band-ratios become less sensitive to changes in Chl-a concentrations because increasing concentrations of color dissolved organic matter (CDOM) and total suspended matter (TSM) (e.g., Bowers et al., 1996) require the use of other spectral bands located in the red (620–700 nm) and NIR (>700 nm) (e.g., Gitelson et al., 2009) (Fig. 5).

Blue–green band-ratios for open and coastal ocean waters

Gordon et al. (1983, 1980) and Feldman et al. (1984) were among the first to use empirical band-ratios from CZCS spectral bands for the study of the near-surface distribution of phytoplankton blooms in the open ocean and to explore their relationships with oceanographic conditions. Many other studies used their initial work to derive global phytoplankton maps from CZCS Chl-a imagery (e.g., Banse and English, 2000, Banse and English, 1997, Banse and English, 1994, Nezlin et al., 1999, Tang et al., 1999 and Fuentes-Yaco et al., 1997b). The second and third generation of ocean color sensors (Fig. 4; Fig. 5; see Fig. 3 of Klemas (2012)) addressed the need for more spectral bands, thereby enabling the development of more sophisticated atmospheric correction schemes and in-water constituent retrieval algorithms, which are required for both improved retrieval accuracy for water quality variables and algal bloom proxies in coastal ocean waters. SeaWiFS OC4 (O'Reilly et al., 2000 and O'Reilly et al., 1998) and MODIS OC3M (Campbell and Feng, 2005b and Campbell and Feng, 2005a) are switching band-ratio algorithms that use spectral bands in the blue and green regions of the visible spectrum to estimate

Chl-a concentrations. The MODIS OC3M (e.g., Chen and Quan, 2013) is extended from the SeaWiFS OC4 and adapted to the MODIS spectral bands. The use of global standard ocean color band-ratios has been demonstrated to significantly overestimate Chl-a. Moore et al. (2009) have shown that the nominal uncertainty of 35% for Chl-a retrievals is true in ocean gyres, but the OC3M relative Chl-a error is >50%outside those gyres and can be >100% in coastal waters. Komick et al. (2009) found that MODIS OC3M systematically overestimated Chl-a when lower than 0.13 mg m^{-3} in Western Canadian waters. Similar results were also found by Radenac et al. (2013) for the equatorial Pacific warm pool. For SeaWiFS OC4, Volpe et al. (2007)found that Chl-a was overestimated by 70% for Chl-a levels lower than 0.2 mg m^{-3} in the Mediterranean Sea. Such low Chl-a concentrations are encountered in the vast majority of the global ocean (Hu et al., 2012).

The relevance of the red-NIR spectral regions in coastal waters

The *in vivo* absorption peak near 676 nm is minimally affected by the influence of CDOM and TSM when the two are in low concentrations (see Section 3.3; Fig. 5). Spectral bands near 676 nm have been widely used for the retrieval of Chl-a in coastal waters (Odermatt et al., 2012 and Gurlin et al., 2011). Gitelson et al. (1999) have shown that reflectance increases in the NIR beyond 700 nm due to increased scattering from algal biomass, correlated to an increase in Chl-a for most phytoplankton groups. The sensitivity analysis conducted by Ruddick et al. (2001) on two red-NIR band-ratio algorithms revealed that the relative error on Chl-a retrievals became more significant at low Chl-a concentrations (<10 mg m^{-3}) and in low backscatter conditions but also that the choice of paired wavelengths was very important. Their study showed that a band-ratio algorithm that uses the red-NIR band pair 672 and 704 nm would perform best at Chl-a ~ 10 mg m^{-3}, whereas a wider red-NIR band pair spreading further apart (e.g., 667 and 748 nm) would perform best at Chl-a ~ 100 mg m^{-3}.

The "red-edge" is technically defined as an increase in spectral reflectance in the red-NIR (680–750 nm) and often results from the presence of partly submersed vegetation (e.g.,Dierssen et al., 2007, Dierssen et al., 2006, Bostater et al., 2003 and Gitelson, 1992) or algal bloom surface expressions (e.g., Shen et al., 2012 and Ruddick et al., 2008) (Fig. 5). Only a few ocean color sensors have the spectral requirements that enable the detection of those reflectance features. For SeaWiFS, two of the nine spectral bands are positioned in red-NIR region of the spectrum (namely 670 nm and 765 nm), and these two bands have limited use for the detection of Chl-a. In contrast, MODIS and MERIS provide more spectral bands between 600 and 800 nm (Fig. 5 and Fig. 6). Further useful applications of the red-NIR spectral region in reflectance-based algorithms are discussed in Sections 3.3 and 4.

Band-ratio algorithms: Limitations and challenges

Most reflectance band-ratios are designed for global applications over optically deep ocean waters (Odermatt et al., 2012). The use of band-ratios often leads to erroneous retrievals in coastal waters, where the optical complexity is highly variable (Dierssen, 2010 and Blondeau-Patissier et al., 2004). Additional limitations in the use of band-ratios are regional differences in optical properties and concentrations; the generalized global parameterization of some algorithms is inapplicable in some of the world's ocean regions (e.g., Volpe et al., 2007, Claustre and Maritorena, 2003, Sathyendranath et al., 2001 and Dierssen and Smith, 2000). Many studies have shown that the retrieval accuracy of Chl-a by satellite ocean color sensors, aimed to be within ±35% in oceanic waters, cannot always be met when using band-ratio algorithms (e.g., Moore et al., 2009 and Hu et al., 2000). In coastal waters, the quality of this retrieval significantly degrades and is often considered unreliable. The use of blue–green spectral bands for the specific detection of Chl-a in coastal waters is affected by the absorption signal of CDOM and TSM (e.g., Dierssen, 2010, Gower, 2000 and Joint and Groom, 2000). To overcome this limitation, other studies suggested the use of red-NIR band-ratios for Chl-a retrieval in coastal waters (e.g., Moses et al., 2012 and Shanmugam, 2011).

Spectral band difference algorithms

Spectral band difference algorithms exploit spectral regions that feature significant changes in the reflectance spectrum due to the presence of an algal bloom, compared to the nearby bloom-free water. Given that absorption tends to vary more rapidly with wavelength than scattering, two adjacent reflectance spectral bands may have similar backscattering properties but will differ significantly in absorption. Hence, this absorption feature can be quantified by spectral difference. The various forms of spectral band difference algorithms use band triplets from the red-NIR or the blue–green spectral regions (Table 2) depending on whether the algorithm is designed to be sensitive to an algal group, high chlorophyll concentrations or surface bloom expressions (Fig. 7). One of the most used ocean color spectral band difference algorithms is the Fluorescence Line Height (FLH) (see review of Xing et al. (2007)), an index for quantifying solar-induced chlorophyll fluorescence. Other similar mathematical expressions are used to derive algal bloom indices from SeaWiFS, MERIS and MODIS (Table 2) and are discussed in this section.

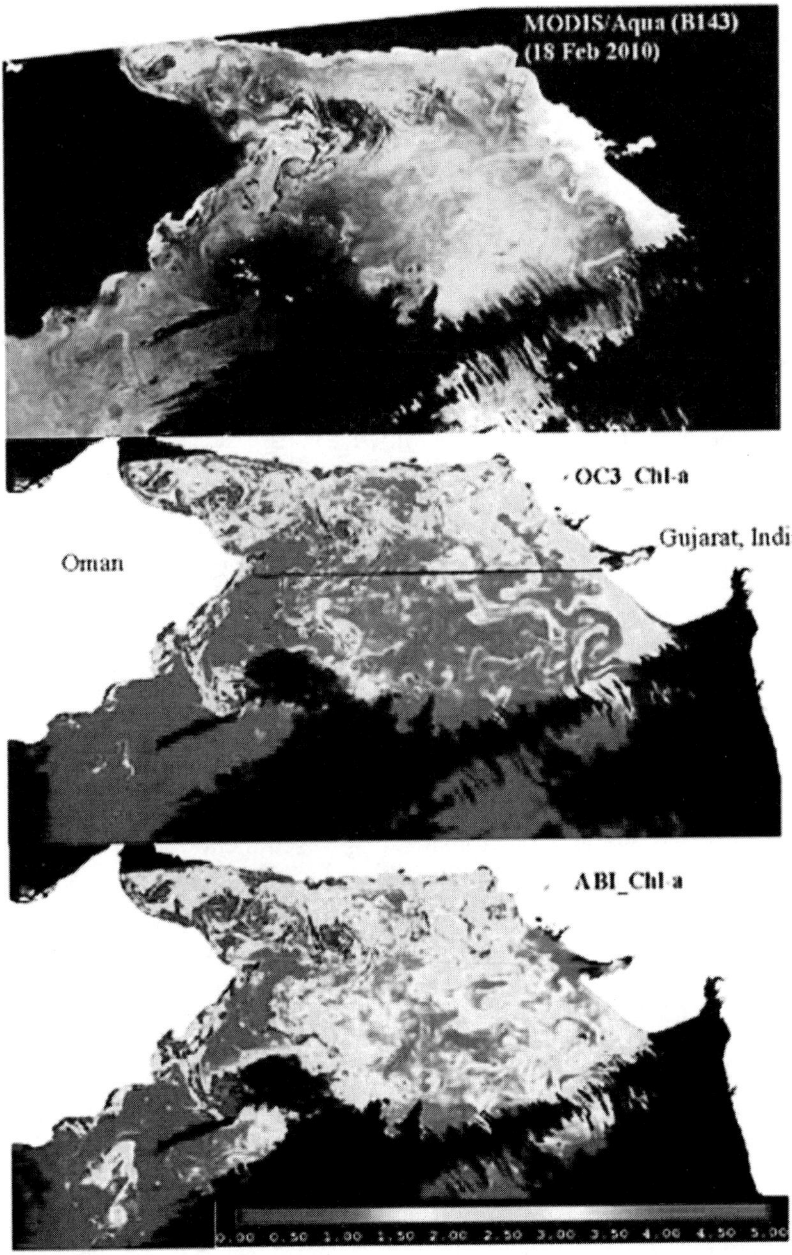

Figure 7. The Algal Bloom Index (ABI) is used to map phytoplankton blooms in the Arabian Sea and the Gulf of Oman from a MODIS-Aqua scene of February 28, 2010. Fig. 4 from Shanmugam (2011). Top panel: A MODIS/Aqua true color composite on 18 February 2010 in the Arabian Sea and Gulf of Oman; The corresponding Chl-a images are derived using (middle) the OC3 and (bottom) ABI algorithms.

Fluorescence Line Height (FLH)

The literature published on this topic since the 1960s (Yentsch and Menzel, 1963) has shown that estimating fluorescence is greatly beneficial to studies of phytoplankton biomass (Falkowski and Kiefer, 1985), physiology (e.g., Westberry et al., 2013 and Behrenfeld et al., 2009), and composition (e.g., Hu et al., 2005). The remote sensing approach used to retrieve FLH was originally developed by Neville and Gower (1977), and its first application to an ocean color sensor was on MODIS-Terra (Abbott and Letelier, 1999 and Letelier and Abbott, 1996). It is well known, however, that MODIS-Terra suffers from uncertainties and instabilities, particularly the radiometric response of the 412-nm band, which has significantly (>40%) degraded since the start of the mission (Franz et al., 2008). The significant striping in MODIS-Terra water-leaving radiances makes the data largely unusable. Thus, MODIS mostly refers to the sensor on the Aqua satellite. The spectral band positions of MERIS (Gower et al., 1999) and MODIS (Hoge et al., 2003) allow for the computation of FLH, but this product cannot be derived from CZCS, SeaWiFS and VIIRS because of the lack spectral bands in the 670–690 nm range. Gower and Borstad (2004) and Zhao et al. (2010) compared FLH results between MODIS and MERIS and concluded that MERIS bands were better positioned for measuring fluorescence. The use of MODIS FLH to detect HAB has been successfully used by many (Frolov et al., 2013, Cannizzaro et al., 2008 and Hu et al., 2005), providing more reliable information than a standard Chl algorithm. MERIS FLH was found to be successful at detecting high biomass phytoplankton in sediment-dominated coastal waters (e.g., Gower et al., 2005). However, others studies have led to inconclusive results on the benefits of FLH in the detection of algal blooms (e.g., Tomlinson et al., 2008).

Maximum Chlorophyll Index (MCI)

The Maximum Chlorophyll Index (MCI) can only be applied to MERIS because of its use of the 708.75 nm band. This band is more responsive to strong reflectance in the NIR, and the lack of similar bands in MODIS and VIIRS may hamper the detection of high-concentration bloom events. The MCI is mainly designed for the detection of high-concentration algal blooms, and it was successfully used to globally monitor phytoplankton blooms in the world's oceans by Gower et al. (2008). The minimum Chl-a concentration required for a phytoplankton bloom to be detected by the MCI is ~30 mg m^{-3} (Gower et al., 2005), but phytoplankton blooms can have much higher concentrations, with some studies reporting Chl-a > 200 mg m^{-3} (e.g., Gower and King, 2007a and Sasamal et al., 2005).

Floating Algae Index (FAI) and Scaled Algae Index (SAI)
Hu et al., 2010c and Hu, 2009 proposed the Floating Algae Index (FAI) to detect large (>4000 km^2) surface-floating algae from MODIS 250 and 500 m bands in both fresh and marine environments. The FAI uses a functional form similar to FLH and MCI where the height of the NIR peak is estimated relatively to a linear baseline from adjacent bands in the red and short wave infrared (SWIR) wavelengths (Table 2). Thus, the FAI is sensitive to the red-edge and is robust to the influence of CDOM, aerosols and sun glint because of the use of NIR bands. However, similarly to MCI, the FAI is likely sensitive to turbid waters and shallow depths. It is used in combination with pre-determined thresholds to help separate land, cloud and high concentrations of submersed algae or sediments from pixels associated with surface algae scums. FAI was originally applied to the detection of cyanobacteria and macro-algae in the freshwater lake Taihu and the Yellow Sea (China). Its global applicability remains untested. Building on this research, Garcia et al. (2013)developed an automatized image processing algorithm, the Scaled Algae Index (SAI), which is a necessary intermediate product for quantifying the spatial coverage of the floating macro-algae observed in satellite imagery based on FAI.

Color Index Algorithm (CIA)
A recent development in algal bloom indices is the Color Index Algorithm (CIA) proposed by Hu et al. (2012). This empirical algorithm was originally developed to estimate surface Chl-a concentrations in oligotrophic ($\leqslant$0.25 mg m^{-3}) waters. The CIA is a three-band reflectance difference algorithm (443 nm, 555 nm and 670 nm bands), making it applicable to SeaWiFS, MODIS and MERIS. Its accuracy has not yet been fully validated because of the lack of low-concentration, high-quality *in situ* Chl-a data from the world's oceans. The CIA was successfully applied to the waters of the Red Sea by Brewin et al. (2013), where it was found to perform better than the MODIS OC3 at retrieving Chl-a because of the low concentrations typically encountered in those waters.

Spectral band difference algorithms: Limitations and challenges
Many limitations in the use of red-NIR bands have been raised in this section. The robust application of FLH for the detection of algal blooms remains under discussion. McKee et al. (2007) suggested that the FLH signal could be masked in complex coastal waters due to the influence of CDOM and TSM. Similarly, Gilerson et al., 2007 and Gilerson et al., 2008 found that MODIS FLH retrieved fluorescence with reasonable accuracy only for waters with Chl < 4 mg m^{-3} but that the signal was masked by particulate backscattering in turbid waters with high CDOM and TSM concentrations. The authors also questioned the performance of

the MERIS FLH algorithm in coastal waters where large errors could be introduced via the linear baseline between the 665 nm and 708.75 nm bands. The use of a linear baseline between the two NIR MERIS bands for the computation of FLH has been demonstrated to work in coastal waters with Chl-a concentrations of up to 20 mg m^{-3} (Gower and King, 2007b) (Fig. 8). For higher chlorophyll concentrations however, due to the combined effects of water absorption and Chl-a distortion, the Chl-a spectrum and a linear baseline can no longer be used. Gower et al. (1999) also challenged the theory that the scattering by TSM had a significant reducing effect on the relative fluorescence height above the baseline. The influence of CDOM on the FLH is often considered small in relatively low concentrations due to its negligible absorption in the NIR. Hu (2009)acknowledged that limitations in the MODIS FAI included the questionable reliability of the MODIS atmospheric correction and the lack of a MODIS-cloud masking algorithm that would reliably flag all cloud and/or sun-glint contaminated pixels while keeping (floating) algae pixels. Although the first issue can be easily solved using Rayleigh-corrected reflectance (Hu et al., 2010b), the second remains a problem, and the author suggested the use of true-color FAI-paired images to separate the clouds from the ocean surface features. Several ocean color algal bloom indices can be combined to enhance our interpretation of algal bloom phenomena and their underlying mechanisms (see Section 5.3).

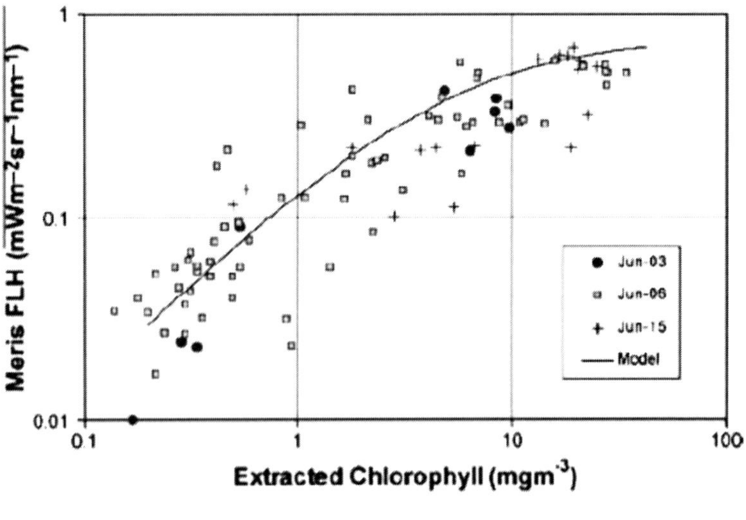

Figure 8. The MERIS FLH product as a function of surface extracted chlorophylls from the Canadian west coast, ECOHAB project, June–September 2003. Fig. 4 of Gower and King (2007b).

Bio-optical models

Bio-optical models are based on various fundamental theories of marine optics and, because they rest on firm theoretical bases and rigorous equations, they are often robust (e.g., Morel, 2001). Coastal systems are generally shallow (<100 m), dynamic, transitional environments that receive considerable amounts of freshwater inputs carrying nutrients, dissolved and particulate organic matter, sediment and contaminants (Blondeau-Patissier et al., 2009 and Babin et al., 2003). Extracting detailed and accurate information from ocean color imagery is a more challenging task in coastal waters in comparison to the open ocean (Tilstone et al., 2011b and Qin et al., 2007). Inversed modeling algorithms (Odermatt et al., 2012) are often necessary to enable the accurate retrieval of Chl-a concentrations in optically complex waters, although optically active constituents (CDOM, TSM) may show discernible patterns of their own during algal bloom conditions (e.g., Chari et al., 2013 and Zhao et al., 2009b).

Retrieval of taxa-specific pigment concentrations from bio-optical models

Taxonomic phytoplankton groups contain different combinations of pigments (e.g., Uitz et al., 2006), but the spectral features of individual pigments are often similar (e.g., Aiken et al., 2008 and Richardson, 1996). There is still some debate about which phytoplankton pigments can be reliably identified because many water quality parameters confine the algorithms to specific conditions. However, the use of the (specific) absorption coefficient of phytoplankton indexed to the composition of the phytoplankton community sampled *in situ* has brought promising results in open ocean waters (e.g., Claustre et al., 2005 and Sathyendranath et al., 2004). Such techniques are employed to detect phytoplankton community structures from ocean color datasets (e.g., Brewin et al., 2011), but significant progress is yet to be made, both in the bio-optical models and databases, for their use in coastal waters.

Derivation of inherent optical properties from bio-optical models for the detection of algal blooms

The principal phytoplankton pigments that mostly contribute to absorption are the photosynthetic pigments consisting of the chlorophylls, carotenoids and biliproteins. The phytoplankton absorption coefficient is an important parameter in ocean color algorithms and is increasingly used to parameterize algal bloom algorithms (e.g., Goela et al., 2013). To study the phytoplankton dynamics characterizing the coastal waters of the Taiwan Strait, Shang et al. (2010) used the Quasi-Analytical Algorithm (QAA) from Lee et al. (2002) to retrieve the phytoplankton absorption coefficient from MODIS. Their findings promoted the use of the phytoplankton absorption coefficient over Chl-a because the latter can suffer from the contamination of non-biotic optically active material.

Bio-optical models: Limitations and challenges
Bio-optical forward and inverse modeling is developing rapidly as an essential tool to understand the effects of dense algal concentrations on light absorption and backscattering coefficients. False positive bloom detections in CDOM- or TSM-rich waters and uncertainties due to the quality of the atmospheric correction are still hampering accurate retrievals of optical properties and biogeochemical concentrations in coastal waters. The use of specific inherent properties now has a central role in bio-optical modeling (e.g., Brando et al., 2012 and Tilstone et al., 2011a), but the identification of specific phytoplankton groups on the basis of inherent optical properties remains a challenge in coastal waters. For future parameterizations of taxa-specific bio-optical models, Claustre et al. (2005) stressed the need to make coincident measurements of the phytoplankton taxonomic composition, photophysiological parameters and phytoplankton absorption to improve local, regional and global *in situ*databases. As suggested by Dierssen (2010) and later by Sen Gupta and McNeil (2012), the water properties of the world's oceans are changing. It is probable that algorithm parameterizations derived from *in situ* data collected over the past decades might not be applicable in the near future due to the possible change in the CDOM–Chl relationships in coastal waters. One of the latest reports on future ocean color mission requirements,IOCCG Report 13 (2012b), stated that "*phytoplankton blooms' timing, frequency, composition and intensity are expected to change with climate in ways that may be hard to predict. (...). Thus reliable and accurate detection of algal blooms is an objective for future missions*" (p. 14).

THE DETECTION OF SPECIFIC TYPES OF ALGAL BLOOMS

Algal blooms with surface expressions

Algal blooms with surface expressions, such as occolithophores, cyanobacteria and*Sargassum*, are observable in satellite imagery due to the large areas they often cover (Fig. 1). Spectral characteristics specific to those taxa may be exploited for their mapping, or masking, by using empirically derived, rule-based reflectance classification algorithms; bio-optical models can also be used.

Coccolithophore blooms

Coccolithophores, with the globally dominant *Emiliania huxleyi*, are calcifying phytoplankton species that form large, dense blooms occurring at most latitudes (Brown, 1995). Coccolithophore blooms play an important role in ocean biogeochemistry (Harlay et al., 2011) and the upper ocean light field (e.g., Balch et al., 2005). The senescent stage of Coccolithophore blooms is often identifiable in an ocean color remote sensing image by its milky blue–green color (Fig. 1), accompanied by the typical high scattering across all spectral bands (400–800 nm) that results from the detached coccoliths (Voss et al., 1998). Following the early work of Holligan et al. (1983), who reported on the higher reflectance values found for CZCS satellite image pixels within Coccolithophore-rich waters, Brown and Yoder (1994a) and Brown (1995) pioneered research on the detection of Coccolithophore blooms. Using a reflectance classification technique to map those events in the global ocean, Brown and Yoder, 1994a and Ackleson et al., 1994 and, later,Merico et al. (2003) identified many limitations that were associated with this technique, including false positives resulting from various sources, such as poor satellite coverage, atmospheric correction errors, reflectance signal contamination from CDOM and scattering from resuspension of TSM. These contaminating sources, whether individual or combined, could bias the results obtained not only for the detection of Coccolithophore blooms but also those of Chl-a (e.g., Blondeau-Patissier et al., 2004). Nonetheless, the acceptable accuracy of the classification results led to the documentation of Coccolithophore bloom events at both global (e.g., Brown, 1995) and basin scales (e.g.,Brown and Podesta, 1997 and Brown and Yoder, 1994b). Since this early work with CZCS (Table 3), novel methods have been developed using reflectance anomalies (e.g.,Shutler et al., 2010) and reflectance classifiers (e.g., Moore et al., 2012) with SeaWiFS (Table 4), MODIS (Table 5) and MERIS (Table 6). These new approaches allowed for the mapping of Coccolithophore blooms in ocean regions where previously such blooms were either not detected or underestimated, such as shelf or polar seas (e.g., Holligan et al., 2010 and Hegseth and Sundfjord, 2008). The use of existing standard Coccolithophore masks in those regions is often unsuitable, and some regional adjustments are required. Iida et al. (2002) demonstrated that the thresholds used in the NASA standard SeaWiFS Coccolithophore mask algorithm were too high to map the events in the Bering Sea and lower threshold values were necessary.

Trichodesmium blooms

Blooms of cyanobacterium *Trichodesmium* fix atmospheric nitrogen into ammonium, making it usable for other organisms (e.g., LaRoche and Breitbarth, 2005). Subramaniam and Carpenter (1994) qualitatively mapped *Trichodesmium sp.* blooms from CZCS satellite imagery of the Atlantic, Indian and Pacific Oceans using an empirical reflectance algorithm based on their high reflectivity. Due to the strong absorption by phycoerythrin,*Trichodesmium* blooms have a characteristic spectral feature at 550 nm that makes them potentially identifiable in satellite reflectance spectra. Bio-optical properties specific to*Trichodesmium* were also used by Subramaniam et al., 1999a and Subramaniam et al., 1999b for the large-scale detection of such events in AVHRR imagery off the Somali coast from a reflectance-based model. Those properties were subsequently used in a semi-empirical bio-optical model by Subramaniam et al. (2002) to detect *Trichodesmium*events in SeaWiFS imagery off the South Atlantic Bight, demonstrating that*Trichodesmium* blooms could be detected in optically complex coastal waters when in sufficient quantity. However, this model had some limitations, including sensitivity to bottom reflection and corals and generating false positives from other similarly reflective sources (such as Coccolithophore or *Phaeocystis* blooms, TSM). Recent studies employ criterion-based reflectance techniques to detect and quantify *Trichodesmium* floating mats using reflectance anomalies (Dupouy et al., 2011 and Dupouy et al., 2008), high-resolution NIR bands (McKinna et al., 2011) and blue–green reflectance bands with MODIS. Hu et al. (2010a) demonstrated the potential of MODIS FAI to distinguish*Trichodesmium* mats from the background influence of both TSM and CDOM in the coastal waters of Florida, on the condition that *Trichodesmium* was in sufficiently high concentration (i.e., Chl-a > 4 mg m^{-3}). Mostly a qualitative approach (e.g., presence/absence), very few of these reflectance-based studies provide quantitative estimates of *Trichodesmium* blooms from satellite imagery. The quantitative estimate of*Trichodesmium sp.* abundances can be achieved with the use of taxa-specific pigment retrieval models. Westberry et al. (2005) used an extension of the bio-optical model ofGarver and Siegel (1997) and Maritorena et al. (2002) (GSM) to map the abundance and biomass of *Trichodesmium*on a global scale from SeaWiFS imagery. These two studies helped reveal many aspects of the distribution of *Trichodemsium sp.* in the global ocean, with the largest annual spatio-temporal occurrence of *Trichodesmium sp.* in the Pacific Ocean.

Floating Sargassum

Sargassum sp. are surface-floating macro-algae distributed throughout the temperate and tropical oceans. *Sargassum* sometimes cover large areas (tens of kilometers), which may provide a useful proxy for tracking convergent or divergent oceanographic processes (e.g., Zhong et al., 2012). MERIS MCI and MODIS FLH have been shown to be very effective in monitoring their distributions in the Gulf of Mexico and the North Atlantic Ocean (Gower and King, 2011b, Gower and King, 2008 and Gower et al., 2006). Using those surface expression indices, Gower and King (2011b) have shown that*Sargassum sp.* seasonal occurrences are characterized by a migration from the Gulf of Mexico in spring to the North Atlantic in July, ending near the Bahamas in February of the following year. An estimated biomass of floating *Sargassum* of 1 g Chl kg^{-1} wet algae (Gower et al., 2006) corresponds to 10^6 ton yr^{-1} sourced from the Gulf of Mexico, which may have large implications for carbon fluxes (Gower and King, 2011b). The spatial and seasonal distributions of *Sargassum* have, overall, been well mapped using satellite imagery, but Gower et al. (2013) recently showed a large drift in their distributions for the year 2011, the cause of which is unclear.

Harmful Algal blooms – Example of dinoflagellate Karenia brevis

The outbreak of a single, dominating phytoplankton species can alter coastal ecosystems, often causing costly damages to both the local economy and the marine ecosystem (e.g., Babin et al., 2008 and Hu et al., 2004). Harmful algal blooms are often referred to as red tide events, but a reddish appearance of the ocean water can be caused by any phytoplankton species, with the only variable being its concentration (Kutser, 2009 and Dierssen et al., 2006). Red-tide-type phytoplankton species, such as dinoflagellate *Karenia brevis*, absorb radiations in the blue and lower green regions (450–500 nm) of the visible spectrum while strongly reflecting radiations in the yellow region (570–580 nm). Bernard et al. (2005) used an empirical Chl-a algorithm based on the ratio between the MERIS red bands 665 and 709 nm to detect high biomass (up to 200 mg m^{-3}) dinoflagellate-dominated HAB events in the Southern Benguela. Another band-ratio was proposed by Ahn and Shanmugam (2006). The Red Tide Index (RI) is specifically designed to detect red-tide blooms (Table 2). The Red Tide Index Chlorophyll Algorithm (RCA) empirically relates RI to Chl-a for concentrations of up to 70 mg m^{-3} (see Fig. 6 from Ahn and Shanmugam (2006)). Their findings showed that spatial patterns of red-tide blooms as mapped from the RI and RCA indices were more consistent with field observations than when standard band-ratio algorithms were used. In this case, Ahn and Shanmugam's (2006) study demonstrated the successful application of the band-ratio approach over bio-optical models such as the one developed by Cannizzaro et al. (2008) for the Gulf of Mexico from SeaWiFS and MODIS imagery. Band difference algorithms are also used. The Red Band

Difference (RBD) proposed by Amin et al., 2009a and Amin et al., 2009b is an index specifically designed to detect *K. brevis*. The spectral region used for the RBD (665–681 nm) is sensitive to both phytoplankton absorption and scattering from suspended sediment; thus, an additional discrimination algorithm is required, the *Karenia brevis* Bloom index (KBBI) (Amin et al., 2008; Table 2). KBBI is the ratio of the RBD to the sum of the same two normalized water reflectance bands. This procedure can be applied to both MODIS and MERIS and has been demonstrated to work in waters other than the Gulf of Mexico (e.g., Gulf Stream, mid-Atlantic).

STATISTICAL TECHNIQUES AND DATA ASSIMILATION TO ASSESS PHYTOPLANKTON BLOOM DYNAMICS

The mapping of the seasonal, inter-annual, or decadal cycle of phytoplankton growth, which can include its spatial patterns, timing and magnitude, is key to understanding algal bloom dynamics in a specific ocean region. The sole use of an ocean color dataset is often insufficient for accurately resolving ocean color constituents (e.g., Mélin et al., 2011, Zingone et al., 2010 and Uz, 2007), and additional data sources from models (e.g.,Mouw et al., 2012) or non-ocean color satellite sensors (e.g., Robinson et al., 2004 and Saitoh et al., 2002)can be necessary. In addition, statistical techniques are often used as a post-processing step to unravel the temporal and/or spatial patterns of those events (e.g., Kurekin et al., 2014). The comprehensive reviews of Bierman et al. (2011) and Kitsiou and Karydis (2011) already described the univariate and multivariate statistical techniques that can be applied to water quality data. Klemas (2012) and Brody et al. (2013) used case studies to illustrate the applications of remote sensing techniques for the detection of phytoplankton blooms. This section provides additional material to these reviews and explores the various statistical techniques that can be employed to extract further information from satellite-derived climatologies and time-series.

Statistical partitioning of marine ecosystems

To analyze the dynamics of phytoplankton blooms in a study region, it is often necessary to first partition it into sub-regions (IOCCG Report 9, 2009). The fundamental difference between the partition of terrestrial and aquatic ecosystems is that aquatic ecosystems usually have dynamic boundaries between biomes to account for seasonal or inter-annual variations (Platt and Sathyendranath, 1999). Longhurst (1995) was among the first to provide a classification of the oceanic pelagic environment based on the spatial variability of the ocean's physical properties. Other work has followed using satellite-derived Chl-a as the main variable to characterize the spatial distribution, areal extent and dynamics of phytoplankton blooms using multivariate statistical analysis (e.g.,Brickley and Thomas, 2004 and Fuentes-Yaco et al., 1997a). Hardman-Mountford et al. (2008) used six years of SeaWiFS Chl monthly global composites and Principal Component Analysis (PCA) to characterize broad-scale ecological patterns in the world's oceans. Their classification, based on the Chl spatial distribution and variability, was consistent with Longhurst's grouping of four primary biomes, although a major difference emerged for the equatorial biome. The delineation of ecosystem boundaries can be performed using clustering methods (e.g., K-means and Empirical Orthogonal Functions) to describe the spatial and temporal variability of a study region (e.g.,Bergamino et al., 2010, Henson and Thomas, 2007 and Brickley and Thomas, 2004) (Fig. 9).

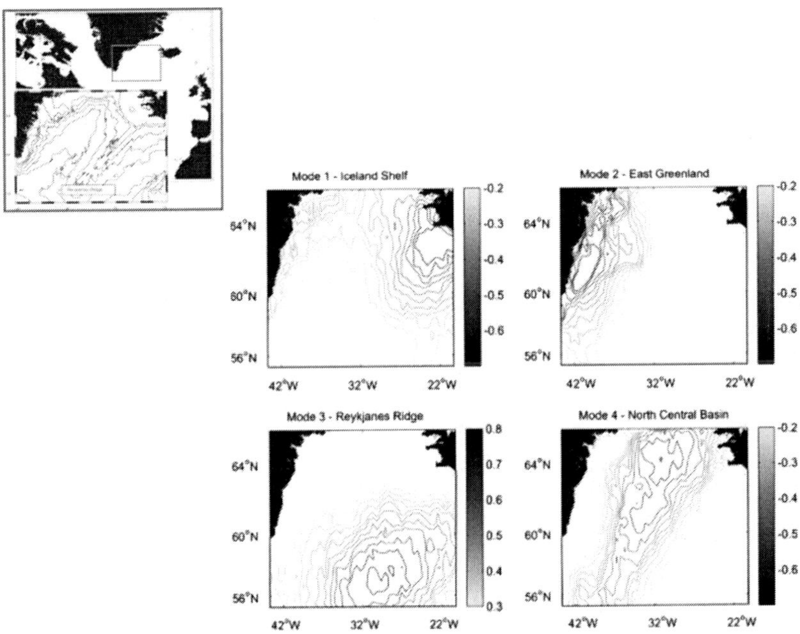

Figure 9. Use of multivariate statistical analysis and satellite ocean color data to define biogeographic zones. The biogeographic zones of the Irminger Basin (top panel) are determined by an EOF analysis of the SeaWiFS chlorophyll-a concentration. The four modes are presented as homogenous correlation maps. Only contours for which the correlation is statistically significant ($p < 0.01$) are plotted. Figs. 1 and 2 fromHenson et al. (2006).

Time-series, fitted models and signal processing techniques

The statistical analysis of algal biomass cycles using time-series can help detect trends in a dataset or reveal correlations between the dynamics of phytoplankton biomass and environmental factors that may be inherent to a study region.

A special issue of the journal Estuaries and Coasts (Zingone et al., 2010) was dedicated to the analysis of multi-scale phytoplankton variability for 22 coastal ecosystems around the world by means of 84 timeseries, some of which were computed from remote sensing data. This special issue highlighted the power of time-series analysis as a statistical technique for deriving environmental baselines. Although SeaWiFS, MODIS-Aqua and MERIS individually have at least a decade of ocean color data, multiple-sensor data merging is often required to significantly extend the timeseries. Maritorena et al. (2010)showed that over a seven year-period, the average daily coverage of the world's ocean was ~25% when a merged satellite product was used, almost double the global coverage of any single mission. Kahru et al. (2010) merged data from SeaWiFS, MERIS and MODIS-Aqua to examine the timing of the annual phytoplankton bloom maximum in the Arctic. The data merging of satellite-derived surface Chl-a extended their time series as far back as 1997 (Table 1 and Table 5). The authors found significant trends towards earlier phytoplankton blooms in ~11% of their study area.

The analysis of timeseries can be approached using time-domain methods or mixed time-and frequency-domain methods. The first technique analyzes data series in the same space in which they were observed, whereas the second decomposes data series at different time scales and frequencies. The sole use of frequency-domain approaches for time-series analyses of phytoplankton biomass is not common. In the timedomain, the timeseries is decomposed into individual, periodic oscillations, the sum of which matches the original signal. Platt et al. (2009) and Song et al. (2010) qualitatively analyzed the timeseries of SeaWiFS data of Chl-a for 10 regions of the Northwest Atlantic Ocean over a 10-year period (1998–2007) with a time-domain approach. To quantify the timing of the spring and summer blooms, they fitted a shifted-Gaussian model to the satellite-derived chlorophyll data using a non-linear least-squares method. They found that blooms in some regions occurred consistently earlier than their expected latitudinal norm.

Mixed time–frequency statistical techniques include the Empirical Mode Decomposition (EMD) (Huang et al., 1998), a signal processing technique that is datadriven and decomposes an initial signal into several high- and low-frequency oscillation components, or wavelets (e.g., Demarcq et al., 2012 and Nezlin and Li, 2003). These components are used to reveal temporal features in noisy timeseries. Wavelet- and variance-preserving spectra, as well as Lomb-Scargle periodograms (Scargle, 1982), are similar to the Fourier Transform. They are used for the detection of cyclic periods of unevenly spaced datasets (e.g., Yoo et al., 2008), which typically include those from ocean color satellites because data gaps can originate from the lack of ocean color sensors (Fig. 4) or from the presence of clouds.

As ocean color satellite data time-series become longer (>10 years of continuous data), time-series analysis techniques will make it possible to more reliably assess the periodicity characterizing the datasets of a specific region. As such, trends in satellite-derived information on phytoplankton are used as a link to possible climate and ecosystem changes.

Satellite product climatologies and merging data from multiple sources

Satellite data climatologies are commonly derived from a three-dimensional datacube (latitude, longitude, time) of a satellite product, allowing the analysis of the selected product through both time and space. Each layer of the datacube is mapped over the same longitude–latitude grid and can be composed of daily satellite images, or composites of various lengths. The use of chlorophyll climatology maps alone can already provide significant information on the spatio-temporal heterogeneity of phytoplankton biomass for a specific region. Yoo et al. (2008) compiled eight years of SeaWiFS Chl data to examine seasonal, inter-annual and event-scale variation in phytoplankton in the North Pacific Ocean. They found that the seasonal progression of the timing of the annual Chl peak showed a different pattern when compared with the Atlantic or Indian Ocean, particularly for the North Pacific High Nutrient–Low Chlorophyll (HNLC) regions, which had distinctive autumnal Chl peaks. HNLC regions comprise the eastern equatorial Pacific Ocean, the sub-arctic North Pacific Ocean, and the Southern Ocean. All three regions combined represent a significant portion (30%) of the global oceanic waters and are typically characterized by low spatio-temporal chlorophyll variability with concentrations <0.3 mg m^{-3} (Boyd et al., 2004). However, higher chlorophyll levels can be observed over the generally shallower continental shelves (Tyrrell et al., 2005).

Several phytoplankton bloom detection techniques make use of satellite product climatologies. Some derive concentration thresholds (Park et al., 2010b and Siegel et al., 2002) for a specific region or pixel. Others involve the use of background subtraction methods (Miller et al., 2006 and Stumpf et al., 2003) or space–time plots (e.g., Quartly and Srokosz, 2004). Despite being effective techniques for flagging bloom anomalies, possible bias can arise from distorted images that may have been unscreened in the compiled climatology. For instance, the uncorrected influence of the sea bottom in shallow clear waters (Barnes et al., 2013) or the persistence of a bloom for a longer period than the timewindow itself may skew the results obtained from those techniques.

The combined use of several remote sensing products can bring a wealth of information that will help understand the environmental factors that may trigger the onset or the termination of algal bloom events in a specific region (Fig. 10 and Fig. 11). Satellite Sea Surface Temperature (SST) data are often used in combination with chlorophyll to relate bloom events to mixed layer depths (Villareal et al., 2012) or upwelling zones (e.g.,Thomas et al., 2012 and Shi and Wang, 2007). In many cases, however, the sole addition of SST does not suffice; Peñaflor et al. (2007) examined the timing and position of a recurrent, seasonal phytoplankton bloom in the Luzon Strait off the Philippines. Using MODIS Chl and SST, in addition to QuickSCAT wind data and bathymetry, they investigated the possible driving forces behind the bloom occurrence. Their study revealed a complex physical–biological interaction for this area and concluded that high phytoplankton biomass resulted from a combination of intense currents, winds and, possibly, freshwater inputs from nearby rivers, rather than upwelling alone, as previously thought. Other recent studies were unable to detect any long-term trend that could characterize phytoplankton dynamics in the world's oceans from the analysis of global, decadal, multi-source ocean color satellite datasets. Martinez et al. (2009) used global Chl datasets from CZCS and SeaWiFS between 1979 and 2002 and concluded that mutlidecadal Chl changes appear to be related to "*basin-scale oscillations in the physical environment. These oscillations can alternately weaken or emphasize the possible effects of global warming, making difficult the identification of trends over short time series*" (p. 1254). Some studies reported an increase in the magnitude of phytoplankton blooms at a regional (e.g., Kahru and Mitchell, 2008 and Richardson and Schoemann, 2004) or global scale (Antoine et al., 2005), whereas others claimed an overall long-term decline in phytoplankton abundance (e.g., Boyce et al., 2010). The latest development in this research is the detection and monitoring of ocean phytoplankton bloom seasonality or phenology (e.g., Demarcq et al., 2012, Racault et al., 2012 and Kahru et al., 2010). An example of phytoplankton bloom phenology at a global scale derived from 10 years of SeaWiFS data is shown in Fig. 12.

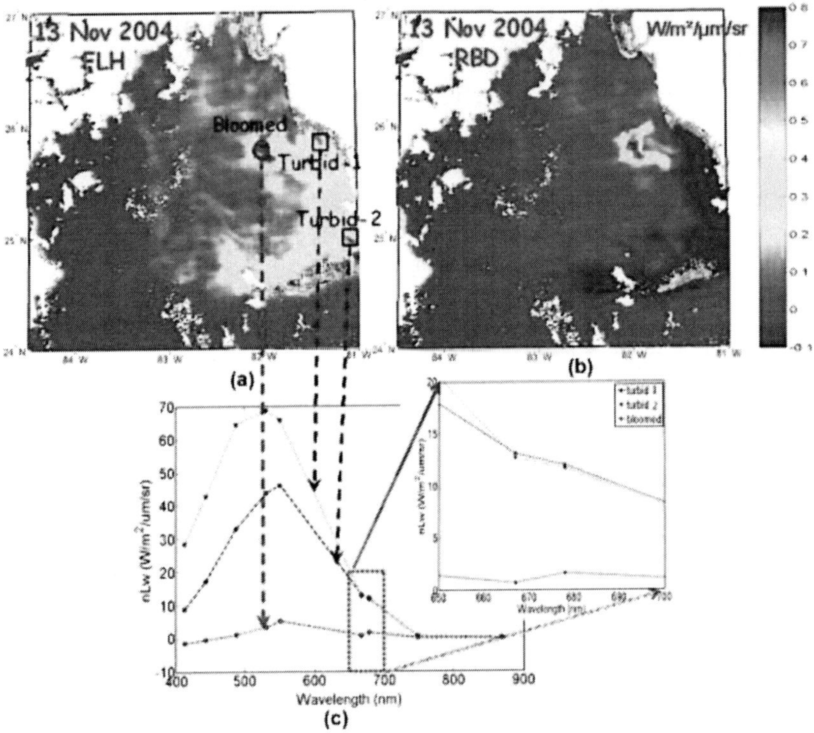

Figure 10. Detecting *Karenia brevis* blooms in a MODIS-Aqua image of November 13, 2004 in west Florida shelf waters using FLH and RBD (both in W m^{-2} μm^{-1} sr^{-1}) (Amin et al. (2009b), Fig. 6).

Data assimilation can be used either for parameter optimization or for state, or flux, estimation. In the context of remote sensing, data assimilation often refers to satellite data being combined with ecosystem models. Zhao et al. (2009a) used several satellite products along with modeled surface current data and geostrophic velocities to explain the occurrence of a near-shore and an offshore phytoplankton bloom in the China Sea following a typhoon. The use of these complementary datasets helped the authors conclude that the near-shore bloom event was induced by terrestrial inputs, whereas the offshore bloom event was triggered by wind-induced upwelling of nutrients. Because ocean color data provide information about water quality variables only (i.e., Chl, TSM, CDOM), they may be limited in their ability to constrain some ecosystem model parameters. This limitation, however, is offset by the volume of data available, covering the full range of biogeochemical responses to the widely varying physical conditions (e.g., Kidston et al., 2013 and Wan et al., 2013). Fully operational algal bloom prediction systems using satellite–ecosystem model data assimilation do exist (e.g., the NOAA

Harmful Algal Bloom Operational Forecast System (HAB-OFS) for the Gulf of Mexico).Carvalho et al. (2011) provided a comparison study that reported on the sensitivity and effectiveness of empirical (e.g., Anderson et al., 2011), bio-optical (Cannizzaro et al., 2008) and operational (e.g., Banks et al. (2012)) methods to detect HAB. The operational method was found to have an elevated frequency of false-negative cases, whereas both the empirical and bio-optical approaches performed similarly, being equally specific and sensitive.

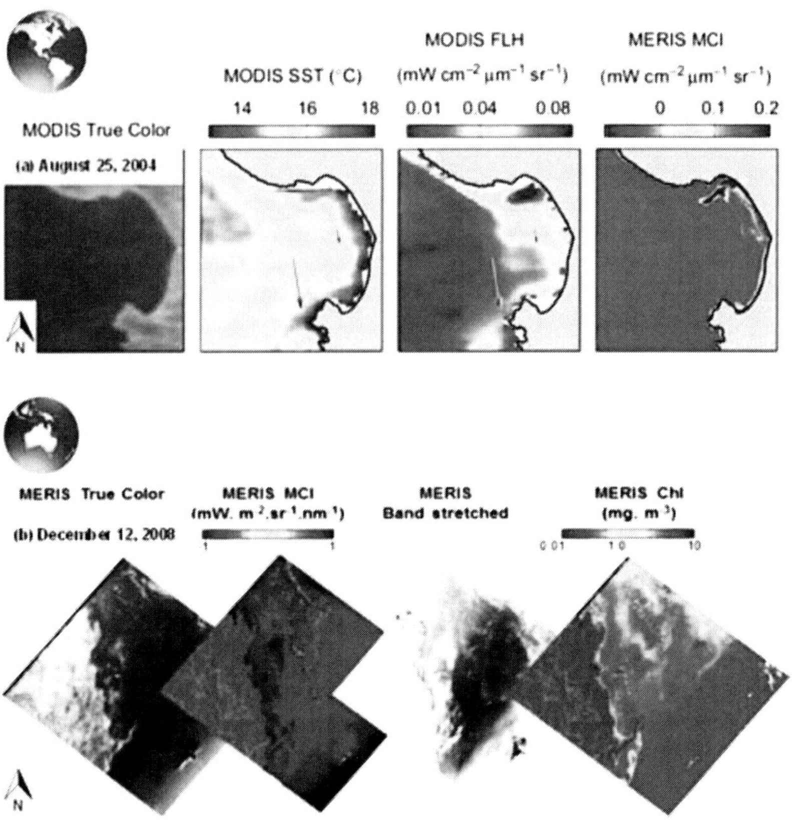

Figure 11. Combining multiple satellite products to identify algal blooms: Top panel: Detecting a red-tide bloom in Monterey Bay, California on August 25, 2004 from MODIS and MERIS products (Fig. 4 from Ryan et al. (2009)); Bottom panel: detecting a surface algal bloom (~500 km long), possibly of *Trichodesmium sp.*, off the Fitzroy River on December 12, 2008 (Great Barrier Reef, South east coast of Australia; 22–26°S) from two MERIS satellite products and a band-stretched image (Blondeau-Patissier (2011)). (For interpretation of the references to colour in this figure legend, the reader is referred to the web version of this article.)

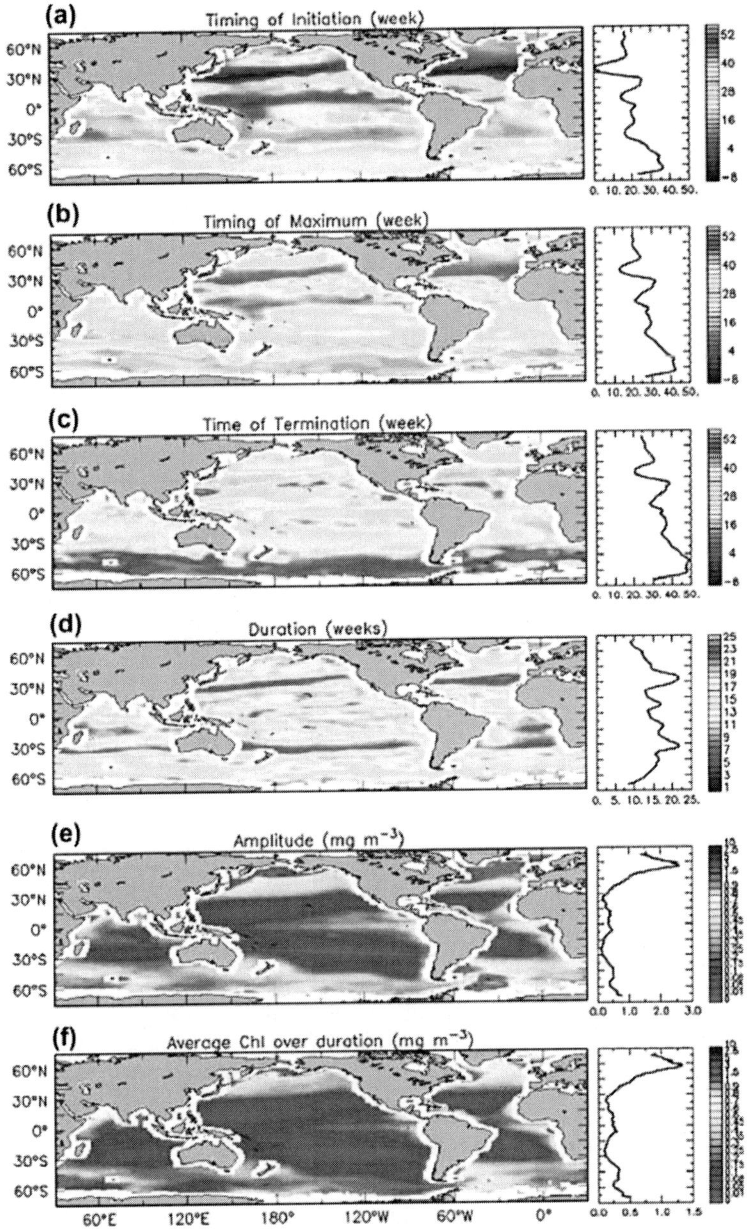

Figure 12. Seasonal timing of phytoplankton on a global scale. The spatial distributions are averaged over 10 years of SeaWiFS data (1998–2007). The right panels show the data averaged longitudinally and smoothed latitudinally with a 5° running average. The negative sign in panels (a–c) is due to the growing period spanning the calendar year. Fig. 1 from Racault et al. (2012).

CONCLUSIONS AND FUTURE DIRECTIONS

We have shown that a wide variety of operational ocean color algorithms currently exist to assist in the detection of phytoplankton blooms in most ocean and coastal regions. For simplistic approaches, such as the sole use of the reflectance spectrum, it is recommended that knowledge of the study area and a further detailed analysis of the pixels' reflectance spectra are taken into account to ensure the validity of the algal bloom information retrieved. Band-ratio algorithms are shown to be adequate for open ocean waters, but their use in complex coastal waters is limited, particularly when band-ratios use blue and green bands because the influence of CDOM and TSM at those wavelengths affects their retrievals. Alternatively, many studies have shown the great potential of band-ratios using red and NIR bands to detect Chl-a and algal blooms in coastal waters (e.g., Le et al., 2013) because this spectral region is less affected by those two optically active constituents. Spectral band difference ocean color indices have also been shown to provide reliable information on algal blooms in both open and coastal ocean waters. Ocean color indices are often used in combination with other descriptors to enhance the interpretation of satellite imagery that possibly features algal bloom events (e.g., Yuan et al., 2005). The MERIS MCI in particular is "*a versatile tool*" (Binding et al., 2012 and Binding et al., 2013), but the FLH and FAI, as well as other indices, can be used in conjunction with satellite estimates of Chl-a and SST to improve our image analysis in a very efficient way.

Significant progress has been made in the use and parameterization of bio-optical models in coastal waters, with a new tendency towards using specific inherent optical properties for the detection of phytoplankton blooms instead of the historically used Chl-a proxy (e.g., Tiwari and Shanmugam, 2013). Assessing the accuracy of bio-optical algorithms for the detection, mapping and monitoring of phytoplankton blooms in coastal waters is a prerequisite for the future. There are currently few optical and biogeochemical*in situ* data characterizing pre- and post-blooms conditions. Automated *in situ* sensors, such as Autonomous Underwater Vehicles (AUV) equipped with bio-optical sensors, may provide a solution (IOCCG Report 11, 2011). In addition to parameters derived from ocean color, today's satellite-derived variables also include Photosynthetic Active Radiations (PAR), wind speed and direction, rainfall, salinity and sea surface height (SSH). These variables can all be combined to provide a more complete assessment of the underlying factors of algal bloom events (e.g., Srokosz et al., 2004 and Urquhart et al., 2013). The use of multi-source datasets from *in situ* data, satellite products or ecosystem models and their analysis with statistical methods is a prerequisite for fully understanding algal blooms' onset mechanisms and dynamics.

Looking back to the potential of ocean color remote sensing, Cracknell et al. (2001) noted that "*operational real-time monitoring of the location, extent, movement and growth rate of a phytoplankton bloom is an important challenge at present*" (p. 221). The capabilities of ocean color remote sensing in providing operational real-time monitoring of a phytoplankton bloom are progressing but continue to remain a challenge in coastal waters (e.g., Malone, 2008). Coastal ocean covers less than 10% of the global ocean surface but accounts for approximately one third of all marine biological productivity (e.g.,Burke et al., 2001). As Siegel et al. (2013) reminded the ocean color community, the future of remote sensing lies in our ability to plan beyond a single sensor mission and to maintain long-term, high-quality, traceable satellite reflectance measurements with sensors' vicarious calibrations, which will make the use of reliable multi-decadal datasets possible. Future satellite instruments and missions should work towards coordinated global coverage of the oceans. Common sets of spectral bands among sensors and a tighter coupling between missions, calibration and validation exercises are desirable (IOCCG Report 13, 2012b). Ocean color satellite mission requirements must be re-evaluated and take into account aspects other than the instrument itself, such as common data processing (e.g., software packages) and management (e.g., centralized data portals). More specifically related to algal blooms, the objectives of future missions are to provide accurate monitoring of HAB, algal bloom timings and magnitudes in optically complex coastal waters, together with the assimilation of satellite-retrieved phytoplankton size or taxa into biogeochemical ecosystem models (e.g., Brewin et al., 2012).

Remote sensing ocean color imagery has provided, and still provides, an invaluable source of frequent, synoptic information. Combined satellite datasets from SeaWiFS, MERIS and MODIS ocean color sensors are equivalent to almost two decades of ocean color imagery for the open and coastal ocean from 1997 to the present, with a brief window of earlier data provided by CZCS from 1979 to 1986. The result is an unprecedented global time series of satellite-derived parameters relevant to algal bloom studies. These ocean color datasets allow for the derivation of ecological baselines, which can then be used to detect and anticipate changes in the ocean system's dynamics (e.g., Siegel et al., 2013, Wong et al., 2009 and Smetacek and Cloern, 2008). In addition, archived earth observation data can be used in hindsight to assess prevailing bloom conditions and to identify the biological and physical parameters (e.g., Kahru et al., 1993) that triggered or terminated an algal event. New missions are often delayed, lost, or cancelled, which results in devastating consequences that likely generate satellite data gaps, such as the 10-year gap between CZCS and SeaWiFS (Fig. 4). This lack of data affected the possibility of answering many environmental questions, one of which was whether the volcanic eruption of Mount Pinatubo in the Philippines in 1991 caused large phytoplankton blooms. The 10 cubic kilometers of material ejected by Mount Pinatubo contained trace metals (Gabrielli et al., 2008), especially iron, that were spread by the winds over the world's oceans. These atmospheric depositions are likely to have generated large-scale phytoplankton blooms, but no ocean color satellite records for those events exist.

Hyperspectral ocean color sensors (Chang et al., 2004) have larger sets of wavelengths that can be used for a more detailed analysis of a remotely sensed signal, whether for chlorophyll retrievals or for the determination of phytoplankton taxa (e.g., Craig et al., 2012 and Torrecilla et al., 2011). Geostationary imagers are the new generation of ocean color sensors (IOCCG Report 12, 2012a). By providing ocean color images for a single region several times a day, datasets from geostationary satellites can help resolve the effects of tides and wind events on coastal currents, which are harder to capture with satellite sensors with a daily repeat cycle. The Geostationary Ocean Color Imager (GOCI), the world's first geostationary orbit satellite, was launched in June 27, 2010, acquiring its first image on July 13, 2010. GOCI is designed to monitor the ocean color around the Korean Peninsula with a spatial resolution of 500 m and a temporal resolution of eight images per day.

It is reasonable to predict that hyperspectral satellite sensors with a full set of spectral bands and higher spectral and temporal resolutions (e.g., Lou and Hu, 2014) will enable the development of effective ocean color algorithms and detection schemes for detecting and monitoring phytoplankton blooms in the open and coastal ocean.

ACKNOWLEDGMENTS

This research was supported by the combination of three funding sources: a PhD scholarship from the International Postgraduate Research Scholarships (The University of Queensland), a top-up scholarship from the Wealth from Oceans National Research Flagship Program (CSIRO) and The North Australian Marine Research Alliance (NAMRA) post-doctoral Fellowship. The authors are grateful to Mrs. Sandra Edgar-Grant

(Charles Darwin University), Dr Ian Leiper (Charles Darwin University) and Elsevier's English Language Editing Service for providing editing support for this manuscript. We would like to thank the space agencies ESA and NASA for providing the satellite images used in this review and the two anonymous reviewers, whose comments significantly improved the paper.

REFERENCES

1. Abbott, M.R., Letelier, R., 1999. MODIS Product Mod-20 Chlorophyll Fluorescence. College of Oceanic and Atmospheric Sciences, Oregon State University, Corvallis, Oregon (USA), p. 42.
2. Acker, J. et al., 2008. Remotely-sensed chlorophyll a observations of the northern red sea indicate seasonal variability and influence of coastal reefs. Journal of Marine Systems 69, 191–204.
3. Ackleson, S.G. et al., 1994. Response of water-leaving radiance to particulate calcite and chlorophyll-a concentrations: a model for gulf of maine coccolithophore blooms. Journal of Geophysical Research 99, 7483–7499.
4. Ahn, Y.H., Shanmugam, P., 2006. Detecting the red tide algal blooms from satellite ocean color observations in optically complex Northeast-Asia coastal waters. Remote Sensing of Environment 103, 419–437.
5. Ahn, Y.-H. et al., 2005. Spatial and temporal aspects of phytoplankton blooms in complex ecosystems off the Korean coast from satellite ocean color observations. Ocean Science Journal 40, 67–71.
6. Ahn, Y.-H. et al., 2006. Satellite detection of harmful algal bloom occurrences in Korean waters. Harmful Algae 5, 213–231.
7. Aiken, J., Hardman-Mountford, N.J., Barlow, R., Fishwick, J., Hirata, T., Smyth, T., 2008. Functional links between bioenergetics and bio-optical traits of phytoplankton taxonomic groups: an overarching hypothesis with applications for ocean colour remote sensing. Journal of Plankton Research 30 (2), 165–181.
8. Amin, R. et al., 2008. Detection of Karenia brevis Harmful Algal Blooms in the West Florida Shelf Using Red Bands of MERIS Imagery. IEEE OCEANS 2008. IEEE, Quebec, Canada.
9. Amin, R. et al., 2009a. MODIS and MERIS detection of dinoflagellate blooms using the Rbd technique. In: 2009 Conference on Remote Sensing of the Ocean, Sea Ice, and Large Water Regions. Berliner Congress Centre, Berlin: SPIE.

10. Amin, R. et al., 2009b. Novel optical techniques for detecting and classifying toxic inoflagellate Karenia brevis blooms using satellite imagery. Optics Express 17,9126–9144.
11. Anderson, C.R. et al., 2011. Detecting toxic diatom blooms from ocean color and a regional ocean model. Geophysical Research Letters 38 (4), L04603.
12. Antoine, D. et al., 2005. Bridging ocean color observations of the 1980s and 2000s in search of long-term trends. Journal of Geophysical Research 110 (C6).
13. Babin, M. et al., 2003. Variations in the light absorption coefficients of phytoplankton, nonalgal particles, and dissolved organic matter in coastal waters around Europe. Journal of Geophysical Research 108 (C7).
14. Babin, S.M. et al., 2004. Satellite evidence of hurricane-induced phytoplankton blooms in an oceanic desert. Journal of Geophysical Research 109 (C03043).
15. Babin, M. et al., 2008. Real-Time Coastal Observing Systems for Marine Ecosystem Dynamics and Harmful Algal Blooms: Theory, Instrumentation and Modelling. UNESCO, Paris.
16. Balch, W.M. et al., 2005. Calcium carbonate measurements in the surface global ocean based on moderate-resolution imaging spectroradiometer data. Journal of Geophysical Research 110, 1–21.
17. Banks, A.C. et al., 2012. A satellite ocean color observation operator system for eutrophication assessment in coastal waters. Journal of Marine Systems 94 (Suppl.), S2–S15.
18. Banse, K., English, D.C., 1994. Seasonality of coastal zone color scanner phytoplankton pigment in the offshore oceans. Journal of Geophysical Research 99, 7323–7345.
19. Banse, K., English, D.C., 1997. Near-surface phytoplankton pigment from the coastal zone color scanner in the subantarctic region southeast of New Zealand. Marine Ecology Progress Series 156, 51–66.
20. Banse, K., English, D.C., 1999. Comparing phytoplankton seasonality in the eastern and western subarctic Pacific and the western Bering Sea. Progress in Oceanography 43, 235–288.
21. Banse, K., English, D.C., 2000. Geographical differences in seasonality of CZCSderived phytoplankton pigment in the Arabian Sea for 1978–1986. Deep Sea Research 47, 1623–1677.
22. Barnes, B.B. et al., 2013. MODIS-derived spatio-temporal water clarity patterns in optically shallow Florida keys waters: a new approach to remove bottom contamination. Remote Sensing of Environment 134, 377–391.
23. Behrenfeld, M.J. et al., 2009. Satellite-detected fluorescence reveals global physiology of ocean phytoplankton. Biogeosciences 6, 779–794.

24. Bergamino, N. et al., 2010. Spatio-temporal dynamics of phytoplankton and primary production in Lake Tanganyika using a MODIS based bio-optical time series. Remote Sensing of Environment 114, 772–780.
25. Bernard, S. et al., 2005. The use of MERIS for harmful algal bloom monitoring in the Southern Benguela. In: Lacoste, H. (Ed.), MAVT, vol. 348. ESRIN, Frascati (It.), pp. 1–7.
26. Bierman, P. et al., 2011. A review of methods for analysing spatial and temporal patterns in coastal water quality. Ecological Indicators 11, 103–114.
27. Binding, C.E. et al., 2012. The MERISMci and its potential for satellite detection of winter diatom blooms on partially ice-covered Lake Erie. Journal of Plankton Research 34, 569–573.
28. Binding, C.E. et al., 2013. The MERIS maximum chlorophyll index; its merits and limitations for inland water algal bloom monitoring. Journal of Great Lakes Research.
29. Blondeau-Patissier, D., 2011. Detection and Quantification of Algal Bloom Dynamics in the Great Barrier Reef Lagoonal Waters Using Remote Sensing and Bio-Optics. School of Geography, Planning and Environmental Management. University of Queensland, Brisbane, p. 334.
30. Blondeau-Patissier, D. et al., 2004. Comparison of bio-physical marine products from SeaWiFS, MODIS and a bio-optical model with in situ measurements from Northern European waters. Journal of Optics A: Pure and Applied Optics 6, 875–889.
31. Blondeau-Patissier, D. et al., 2009. Bio-optical variability of the absorption and scattering properties of the Queensland inshore and reef waters, Australia. Journal of Geophysical Research 114 (C05003).
32. Bostater Jr., C.R. et al., 2003. Hyperspectral remote sensing protocol development for submerged aquatic vegetation in shallow waters. In: Bostater, C.R., Jr., Santoleri, R. (Eds.), Remote Sensing of the Ocean and Sea Ice, vol. 5233. SPIE, Barcelona, Spain, pp. 199–215.
33. Bowers, D.G. et al., 1996. Absorption spectra of inorganic particles in the Irish Sea and their relevance to remote sensing of chlorophyll. International Journal of Remote Sensing 17, 2449–2460.
34. Boyce, D.G. et al., 2010. Global phytoplankton decline over the past century. Nature 466, 591–596.
35. Boyd, P.W. et al., 2004. Episodic enhancement of phytoplankton stocks in New Zealand subantarctic waters: contribution of atmospheric and oceanic iron supply. Global Biogeochemical Cycles 18, GB1029.
36. Brando, V.E. et al., 2012. An adaptive semi-analytical inversion of ocean colour radiometry in optically complex waters. Applied Optics 51, 2808–2833.

37. Brewin, R.J.W. et al., 2011. An Intercomparison of bio-optical techniques for detecting dominant phytoplankton size class from satellite remote sensing. Remote Sensing of Environment 115, 325–339.
38. Brewin, R.J.W. et al., 2012. The influence of the Indian Ocean dipole on interannual variations in phytoplankton Size Structure as revealed by earth observation. Deep-Sea Research Part II: Topical Studies in Oceanography 77–80, 117–127.
39. Brewin, R.J.W., Raitsos, D.E., Pradhan, Y., Hoteit, I., 2013. Comparison of chlorophyll in the Red Sea derived from MODIS-Aqua and in vivo fluorescence. Remote Sensing of Environment 136, 218–224.
40. Bricaud, A., Claustre, H., Ras, J., Oubelkheir, K., 2004. Natural variability of phytoplanktonic absorption in oceanic waters: influence of the size structure of algal populations. Journal of Geophysical Research 109, C11010.
41. Brickley, P.J., Thomas, A.C., 2004. Satellite-measured seasonal and inter-annual chlorophyll variability in the Northeast Pacific and coastal Gulf of Alaska. Deep Sea Research Part II: Topical Studies in Oceanography 51, 229–245.
42. Brody, S.R., Lozier, M.S., Dunne, J.P., 2013. A comparison of methods to determine phytoplankton bloom initiation. Journal of Geophysical Research: Oceans 118 (5), 2345–2357.
43. Brown, C.W., 1995. Global distribution of coccolithophore blooms. Oceanography 8, 59–60.
44. Brown, C.W., Podesta, G.P., 1997. Remote sensing of coccolithophore blooms in the western South Atlantic ocean. Remote Sensing of Environment 60, 83–91.
45. Brown, C.W., Yoder, J.A., 1994a. Coccolithophorid blooms in the global ocean. Journal of Geophysical Research 99, 7467–7482.
46. Brown, C.W., Yoder, J.A., 1994b. Distribution pattern of coccolithophorid blooms in the western North Atlantic Ocean. Continental Shelf Research 14, 175–197.
47. Burke, L.A. et al., 2001. Pilot Analysis of Global Ecosystems: Coastal Ecosystems. World Resources Institute, Washington, DC, p. 94.
48. Campbell, J.W., Feng, H., 2005a. The Empirical Chlorophyll Algorithm for MODIS: Testing the Oc3m Algorithm Using Nomad Data. Ocean Color Bio-optical Algorithm Mini Workshop (OCBAM). University of New Hampshire, Durham (USA): Ocean Color Bio-optical Algorithm Mini Workshop (OCBAM).
49. Campbell, J.W., Feng, H., 2005b. The Empirical Chlorophyll Algorithm for SeaWiFS: Testing the Oc4.V4 Algorithm Using Nomad Data Ocean Color Bio-optical Algorithm Mini Workshop (OCBAM). University of New Hampshire, Durham (USA).

50. Cannizzaro, J., 2004. Detection and Quantification of Karenia brevis Blooms on the West Florida Shelf from Remotely Sensed Ocean Color Imagery Vol. MSc College of Marine Science, University of South Florida, p. 81.
51. Cannizzaro, J.P. et al., 2008. A novel technique for detection of the toxic dinoflagellate, Karenia brevis, in the Gulf of Mexico from remotely sensed ocean color data. Continental Shelf Research 28, 137–158.
52. Carvalho, G.A. et al., 2010. Satellite remote sensing of harmful algal blooms: a new multi-algorithm method for detecting the Florida red tide (Karenia brevis). Harmful Algae 9, 440–448.
53. Carvalho, G.A. et al., 2011. Long-term evaluation of three satellite ocean color algorithms for identifying harmful algal blooms (Karenia brevis) along the west coast of Florida: a matchup assessment. Remote Sensing of Environment 115,1–18.
54. Chang, N.B., Xuan, Z.M., 2011. Exploring the nutrient inputs and cycles in Tampa Bay and coastal watersheds using MODIS images and data mining. In: Gao, W.et al. (Eds.), Remote Sensing and Modeling of Ecosystems for Sustainability Viii, vol. 8156. Spie-IntSoc Optical Engineering, Bellingham.
55. Chang, G. et al., 2004. The new age of hyperspectral oceanography. Oceanography, 23–29.
56. Chari, N.V.H.K. et al., 2013. Fluorescence and absorption characteristics of dissolved organic matter excreted by phytoplankton species of Western Bay of Bengal under Axenic Laboratory condition. Journal of Experimental Marine Biology and Ecology 445, 148–155.
57. Chen, J., Quan, W., 2013. An improved algorithm for retrieving chlorophyll-a from the yellow river estuary using MODIS imagery. Environmental Monitoring and Assessment 185, 2243–2255.
58. Claustre, H., Maritorena, S., 2003. The many shades of ocean blue. Science 302, 1514–1515.
59. Claustre, H., Babin, M., Merien, D., Ras, J., Prieur, L., Dallot, S., Prasil, O., Dousova, H., Moutin, T., 2005. Toward a taxon-specific parameterization of bio-optical models of primary production: a case study in the North Atlantic.. Journal of Geophysical Research 110, C07S12.
60. Cracknell, A.P. et al., 2001. The Abdmap (algal bloom detection, monitoring and prediction) concerted action. International Journal of Remote Sensing 22, 205– 247.
61. Craig, S.E. et al., 2012. Deriving optical metrics of coastal phytoplankton biomass from ocean colour. Remote Sensing of Environment 119, 72–83.
62. Cullen, J.J., 1982. The deep chlorophyll maximum: comparing vertical profiles of chlorophyll-A. Canadian Journal of Fisheries and Aquatic Sciences 39, 791–803.

63. Dasgupta, S. et al., 2009. Comparison of global chlorophyll concentrations using MODIS data. Advances in Space Research 43, 1090–1100.
64. Demarcq, H. et al., 2012. Monitoring marine phytoplankton seasonality from space. Remote Sensing of Environment 117, 211–222.
65. Dickey, T. et al., 2006. Optical oceanography: recent advances and future directions using global remote sensing and in situ observations. Reviews of geophysics 44, RG1001.
66. Dierssen, H.M., 2010. Perspectives on empirical approaches for ocean color remote sensing of chlorophyll in a changing climate. Proceeding of the National Academy of Sciences of the United States of America 107, 17073–17078.
67. Dierssen, H.M., Smith, R.C., 2000. Bio-optical properties and remote sensing ocean color algorithms for Antarctic Peninsula waters. Journal of Geophysical Research 105, 26301–26312.
68. Dierssen, H.M. et al., 2006. Red and black tides: quantitative analysis of waterleaving radiance and perceived color for phytoplankton, colored dissolved organic matter, and suspended sediments. Limnology and Oceanography 51, 2646–2659.
69. Dierssen, H.M. et al., 2007. The Red Edge: Exploring High Near-Infrared Reflectance of Phytoplankton and Submerged Macrophytes and Implications for Aquatic Remote Sensing. AGU Spring Meeting 2007: American Geophysical Union. Dore, J.E., Letelier, R.M., Church, M.J., Lukas, R., Karl, D.M., 2008. Summer phytoplankton blooms in the oligotrophic North Pacific Subtropical Gyre: historical perspective and recent observations. Progress in Oceanography 76 (1), 2–38.
70. Dupouy, C. et al., 2008. Bio-optical properties of the marine cyanobacteria Trichodesmium spp. Journal of Applied Remote Sensing 2 (1), 023503.
71. Dupouy, C. et al., 2011. An algorithm for detecting Trichodesmium surface blooms in the south western tropical Pacific. Biogeosciences Discussions 8, 5653–5689.
72. Falkowski, P., Kiefer, D.A., 1985. Chlorophyll-a fluorescence in phytoplankton: relationship to photosynthesis and biomass. Journal of Plankton Research 7, 715–731.
73. Fargion, G.S., McClain, C.R., 2003. An Overview of Simbios Program Activities and Accomplishments. SIMBIOS Project; 2003 Annual Report. NASA Goddard Space Flight Center, Greenbelt (Maryland, USA), pp. 1–33.
74. Feldman, G. et al., 1984. Satellite color observations of the phytoplankton distribution in the eastern equatorial Pacific during the 1982–1983 El Niño. Science 226, 1069–1071.

75. Focardi, S. et al., 2009. A combinated approach to investigate the biochemistry and hydrography of a shallow bay in the South Adriatic Sea: the Gulf of Manfredonia (Italy). Environmental Monitoring and Assessment 153, 209–220.
76. Franz, B., Kwiatowska, E.J., Meister, G., McClain, C.R., 2008. Moderate resolution imaging spectroradiometer on Terra: limitations for ocean color applications. Journal of Applied Remote Sensing 2 (1).
77. Frolov, S. et al., 2013. Monitoring of harmful algal blooms in the era of diminishing resources: a case study of the U.S. West Coast. Harmful Algae 21–22, 1–12.
78. Fuentes-Yaco, C. et al., 1997a. Phytoplankton pigment in the Gulf of St. Lawrence, Canada, as determined by the coastal zone color scanner – Part Ii: Multivariate Analysis. Continental Shelf Research 17, 1441–1459.
79. Fuentes-Yaco, C. et al., 1997b. Phytoplankton pigment in the Gulf of St. Lawrence, Canada, as determined by the coastal zone color scanner – Part I: Spatiotemporal variability. Continental Shelf Research 17, 1421–1439.
80. Gabric, A.J. et al., 1990. Spatio-temporal variability in surface chlorophyll distribution in the central Great Barrier Reef as derived from CZCS imagery. Australian Journal of Marine and Freshwater Research 41, 313–324.
81. Gabrielli, P. et al., 2008. Siderophile metal fallout to Greenland from the 1991 winter eruption of Hekla (Iceland) and during the global atmospheric perturbation of Pinatubo. Chemical Geology 255, 78–86.
82. Garcia, C.A.E., Garcia, V.M.T., 2008. Variability of chlorophyll-a from ocean color images in the La Plata continental shelf region. Continental Shelf Research 28, 1568–1578.
83. Garcia, R. et al., 2013. Quantification of Floating Macroalgae Blooms Using the Scaled Algae Index (Sai). Journal of Great Lakes Research 118, 26–46.
84. Garcia-Soto, C., Pingree, R.D., 2009. Spring and summer blooms of phytoplankton (SeaWiFS/MODIS) along a ferry line in the Bay of Biscay and western English Channel. Continental Shelf Research 29, 1111–1122.
85. Garver, S.A., Siegel, D.A., 1997. Inherent optical property inversion of ocean color spectra and its biogeochemical interpretation – 1. Time series from the Sargasso Sea. Jounal of Geophysical Research 102, 18607–18625.
86. Gazeau, F., Smith, S.V., Gentili, B., Frankignoulle, M., Gattuso, J.-P., 2004. The European coastal zone: characterization and first assessment of ecosystem metabolism. Estuarine, Coastal and Shelf Science 60 (4), 673–694.
87. Gilerson, A. et al., 2007. Fluorescence component in the reflectance spectra from coastal waters. I. Dependence on water composition. Optics Express 15, 15702–15721.

88. Gilerson, A. et al., 2008. Fluorescence component in the reflectance spectra from coastal waters. Ii. Performance of retrieval algorithms. Optics Express 16, 2446–2460.
89. Gitelson, A., 1992. The peak near 700 nm on radiance spectra of algae and water: relationships of its magnitude and position with chlorophyll concentration. International Journal of Remote Sensing 13, 3367–3373.
90. Gitelson, A. et al., 1999. Comparative reflectance properties of algae cultures with manipulated densities. Journal of Applied Phycology 11, 345–354.
91. Gitelson, A.A. et al., 2009. A bio-optical algorithm for the remote estimation of the chlorophyll-a concentration in case 2 waters. Environmental Research Letters 4, 5.
92. Goela, P.C. et al., 2013. Specific absorption coefficient of phytoplankton off the southwest coast of the Iberian Peninsula: a contribution to algorithm development for ocean colour remote sensing. Continental Shelf Research 52, 119–132.
93. Gohin, F. et al., 2003. Satellite and in situ observations of a late winter phytoplankton bloom, in the Northern Bay of Biscay. Continental Shelf Research 23, 1117–1141.
94. Gordoa, A. et al., 2008. Spatio-temporal patterns in the North-Western Mediterranean from MERIS derived chlorophyll-a concentration. Scientia Marina 72, 757–767.
95. Gordon, H.R. et al., 1980. Phytoplankton pigments from the Nimbus-7 coastal zone color scanner: comparisons with surface measurements. Science 210, 63–66.
96. Gordon, H.R. et al., 1983. Phytoplankton pigment concentrations in the Middle Atlantic Bight: comparison of ship determinations and coastal zone color scanner measurements. Applied Optics 22, 4175–4180.
97. Gower, J., 2000. Productivity and plankton blooms observed with SeaWiFS. In: Proceedings of the Fifth Pacific Ocean Remote Sensing Conference (Porsec), 5–8 December 2000, vol. 1. NIO Dona Paula Goa (India), pp. 23–27.
98. Gower, J., 2001. Productivity and plankton blooms observed with SeaWiFS and in situ sensors. IEEE Geoscience and Remote Sensing Letters 5, 2181–2183.
99. Gower, J., Borstad, G.A., 2004. On the potential of MODIS and MERIS for imaging chlorophyll fluorescence from space. International Journal of Remote Sensing 25, 1459–1464.
100. Gower, J., King, S., 2007a. An Antarctic ice-related ‘‘superbloom’’ observed with the MERIS satellite imager. Geophysical Research Letters, 34.

101.Gower, J., King, S., 2007b. Validation of chlorophyll fluorescence derived from MERIS on the west coast of Canada. International Journal of Remote Sensing 28, 625– 635.

102.Gower, J., King, S., 2008. Satellite Images Show the Movement of Floating Sargassum in the Gulf of Mexico and Atlantic Ocean. Nature Proceedings. Gower, J., King, S., 2011a. A global survey of intense surface plankton blooms and floating vegetation using MERIS MCI. In: Tang, D. (Ed.), Remote Sensing of the Changing Oceans. Springer, p. 396.

103.Gower, J., King, S.A., 2011b. Distribution of floating Sargassum in the Gulf of Mexico and the Atlantic Ocean mapped using MERIS. International Journal of Remote Sensing 32, 1917–1929.

104.Gower, J. et al., 1999. Interpretation of the 685 nm peak in water-leaving radiance spectra in terms of fluorescence, absorption and scattering, and its observation by MERIS. International Journal of Remote Sensing 20, 1771–1786.

105.Gower, J. et al., 2003. Use of 709 nm band of MERIS to detect intense plankton blooms and other conditions in coastal waters. In: ESA (Ed.), MERIS User Workshop. ESA, Frascati (It.), p. 6.

106.Gower, J. et al., 2004. Observation of chlorophyll fluorescence in west coast waters of Canada using the MODIS satellite sensor. Canadian Journal of Remote Sensing 30, 17–25.

107.Gower, J. et al., 2005. Detection of intense plankton blooms using the 709 nm band of the MERIS imaging spectrometer. International Journal of Remote Sensing 26, 2005–2012.

108.Gower, J. et al., 2006. Ocean color satellites show extensive lines of floating Sargassum in the Gulf of Mexico. IEEE Transaction on Geoscience and Remote Sensing 44, 3619–3625. http://dx.doi.org/10.1109/TGRS.2006.882258.

109.Gower, J. et al., 2008. Global monitoring of plankton blooms using MERIS MCI. International Journal of Remote Sensing 29, 21.

110.Gower, J., Young, E., King, S., 2013. Satellite images suggest a new Sargassum source region in 2011. Remote Sensing Letters 4 (8), 764–773.

111.Gurlin, D. et al., 2011. Remote estimation of chl-a concentration in turbid productive waters: return to a simple two-band NIR-red model? Remote Sensing of Environment 115, 3479–3490.

112.Hardman-Mountford, N.J. et al., 2008. An objective methodology for the classification of ecological pattern into biomes and provinces for the pelagic ocean. Remote Sensing of Environment 112, 3341–3352.

113.Harlay, J. et al., 2011. Biogeochemical study of a coccolithophore bloom in the Northern Bay of Biscay (Ne Atlantic Ocean) in June 2004. Progress in Oceanography 86, 317–336.

114. Hegseth, E.N., Sundfjord, A., 2008. Intrusion and blooming of Atlantic phytoplankton species in the high arctic. Journal of Marine Systems 74, 108–119.
115. Henson, S.A., Thomas, A.C., 2007. Interannual variability in timing of bloom initiation in the California current system. Journal of Geophysical Research 112, C8.
116. Henson, S.A. et al., 2006. Effect of meteorological conditions on interannual variability in timing and magnitude of the spring bloom in the Irminger Basin, North Atlantic. Deep Sea Research (Part I) 53, 1601–1615.
117. Henson, S.A. et al., 2009. Decadal variability in north Atlantic phytoplankton blooms. Journal of Geophysical Research C: Oceans 114, C04013.
118. Hoge, F.E. et al., 2003. Validation of terra-MODIS phytoplankton chlorophyll fluorescence line height. I. Initial airborne lidar results. Applied Optics 42 (15).
119. Holligan, P.M. et al., 1983. Satellite and ship studies of coccolithophore production along a continental shelf edge. Nature 304, 339–342.
120. Holligan, P.M. et al., 2010. Seasonal distributions of the coccolithophore, Emiliania huxleyi, and of particulate inorganic carbon in surface waters of the Scotia Sea. Journal of Marine Systems 82, 195–205.
121. Hu, C., 2009. A novel ocean color index to detect floating algae in the global oceans. Remote Sensing of Environment 113, 2118–2129.
122. Hu, C. et al., 2000. How precise are SeaWiFS ocean color estimates? Implications of digitization-noise errors. Remote Sensing of Environment 76, 239–249.
123. Hu, C. et al., 2004. Linkages between coastal runoff and the Florida Keys ecosystem: a study of a Dark Plume event. Geophysical Research Letters 31, L15307.
124. Hu, C. et al., 2005. Red tide detection and tracing using MODIS fluorescence data: a regional example in SW Florida coastal waters. Remote Sensing of Environment 97, 311–321.
125. Hu, C. et al., 2010a. Remote detection of Trichodesmium blooms in optically complex coastal waters: examples with MODIS full-spectral data. Remote Sensing of Environment 114, 2048–2058.
126. Hu, C. et al., 2010b. On the recurrent Ulva prolifera blooms in the yellow sea and East China Sea. Journal of Geophysical Research 115, C05017.
127. Hu, C. et al., 2010c. Moderate Resolution Imaging Spectroradiometer (MODIS) observations of cyanobacteria blooms in Taihu Lake, China. Journal of Geophysical Research-Oceans 115, C04002.
128. Hu, C. et al., 2012. Chlorophyll a algorithms for oligotrophic oceans: a novel approach based on three-band reflectance difference. Journal of Geophysical Research 117 (C1), C01011.

129.Huang, N.E. et al., 1998. The Empirical Mode Decomposition and the Hilbert Spectrum for Nonlinear and Non-Stationary Time Series Analysis. Royal Society of London, vol. 454. London, pp. 903–995.

130.Huisman, J., Pham Thi, N.N., Karl, D.M., Sommeijer, B., 2006. Reduced mixing generates oscillations and chaos in the oceanic deep chlorophyll maximum. Nature 439 (7074), 322–325.

131.Iida, T., Saitoh, S.-I., 2007. Temporal and spatial variability of chlorophyll concentrations in the Bering Sea using empirical orthogonal function (Eof) analysis of remote sensing data. Deep Sea Research Part II: Topical Studies in Oceanography 54, 2657–2671.

132.Iida, T. et al., 2002. Temporal and spatial variability of coccolithophore blooms in the eastern Bering Sea, 1998–2001. Progress in Oceanography 55, 165–175.

133.Iida, T. et al., 2012. Interannual variability of coccolithophore Emiliania huxleyi blooms in response to changes in water column stability in the eastern Bering Sea. Continental Shelf Research 34, 7–17.

134.IOCCG, 2008. Why ocean colour? The societal benefits of ocean-colour technology. In: Platt, T. et al. (Eds.), Reports of the International Ocean-Colour Coordinating Group. Bedford Institute, Dartmouth (Nova Scotia, Ca.).

135.IOCCG, 2009. Partition of the ocean into ecological provinces: role of ocean-colour radiometry. In: Dowell, M., Platt, T. (Eds.), Reports of the International OceanColour Coordinating Group. Bedford Institute, Dartmouth (Nova Scotia, Ca.).

136.IOCCG, 2011. Bio-optical sensors on argo floats. In: Claustre, H. (Ed.), Reports of the International Ocean Colour Coordinating Group. Laboratoire d'Océanographie de Villefranche (LOV-CNRS), Villefranche-sur-mer (Fr.), p. 89.

137.IOCCG, 2012a. Ocean-colour observations from a geostationary orbit. In: Antoine, D. (Ed.), Reports of the International Ocean Colour Coordinating Group. Laboratoire d'Océanographie de Villefranche (LOV-CNRS), Villefranche-surmer (Fr.), p. 102.

138.IOCCG, 2012b. Mission requirements for future ocean-colour sensors. In: McClain, C., Meister, G. (Eds.), Reports of the International Ocean Colour Coordinating Group. NASA Goddard Space Flight Center, Greenbelt (MD, USA), p. 106.

139.Jessup, D.A. et al., 2009. Mass stranding of marine birds caused by a surfactantproducing red tide. PLoS One 4, 8.

140.Joint, I.I., Groom, S.B., 2000. Estimation of phytoplankton production from space: current status and future potential of satellite remote sensing. Journal of Experimental Marine Biology and Ecology 250, 233–255.

141.Kahru, M., Mitchell, B.G., 2008. Ocean color reveals increased blooms in various parts of the world. Eos, Transactions, American Geophysical Union 89, 170.

142.Kahru, M. et al., 1993. Cyanobacterial blooms cause heating of the sea surface. Marine Ecology Progress Series 101, 1–7.

143.Kahru, M. et al., 2004. MODIS detects a devastating algal bloom in paracas bay, Peru. Eos, Transactions, American Geophysical Union 85, 465–472.

144.Kahru, M. et al., 2010. Are phytoplankton blooms occurring earlier in the Arctic? Global Change Biology, 1733–1739.

145.Kidston, M., Matear, R., Baird, M.E., 2013. Phytoplankton growth in the Australian sector of the Southern Ocean, examined by optimising ecosystem model parameters. Journal of Marine Systems 128, 123–137.

146.Kim, S. et al., 2000. Temporal and spatial variability of phytoplankton pigment concentrations in the Japan sea derived from CZCS images. Journal of Oceanography 56, 527–538.

147.Kitsiou, D., Karydis, M., 2011. Coastal marine eutrophication assessment: a review on data analysis. Environment International 37, 778–801.

148.Klemas, V., 2011. Remote sensing techniques for studying coastal ecosystems: an overview. Journal of Coastal Research 27, 2–17.

149.Klemas, V., 2012. Remote sensing of algal blooms: an overview with case studies. Journal of Coastal Research 28, 34–43.

150.Kogeler, J., Rey, F., 1999. Ocean colour and the spatial and seasonal distribution of phytoplankton in the Barents Sea. International Journal of Remote Sensing 20, 1303–1318.

151.Komick, N.M. et al., 2009. Bio-optical algorithm evaluation for MODIS for Western Canada coastal waters: an exploratory approach using in situ reflectance. Remote Sensing of Environment 113, 794–804.

152.Kratzer, S. et al., 2013. The use of ocean color remote sensing in integrated coastal zone management—a case study from Himmerfjärden, Sweden. Marine Policy.

153.Kurekin, A.A., Miller, P.I., Van der Woerd, H.J., 2014. Satellite discrimination of Karenia mikimotoi and Phaeocystis harmful algal blooms in European coastal waters: merged classification of ocean colour data. Harmful Algae 31, 163– 176.

154.Kutser, T., 2009. Passive optical remote sensing of cyanobacteria and other intense phytoplankton blooms in coastal and inland waters. International Journal of Remote Sensing 30, 4401–4425.

155.LaRoche, J., Breitbarth, E., 2005. Importance of the diazotrophs as a source of new nitrogen in the ocean. Journal of Sea Research 53, 67–91.

156.Lavender, S.J., Groom, S.B., 2001. The detection and mapping of algal blooms from space. International Journal of Remote Sensing 22, 197–201.

157.Le, C. et al., 2013. Towards a long-term chlorophyll-a data record in a turbid estuary using MODIS observations. Progress in Oceanography 109, 90–103.

158.Lee, Z. et al., 2002. Deriving inherent optical properties from water color: a multiband quasi-analytical algorithm for optically deep waters. Applied Optics 41, 5755–5772.

159.Lee, Z., Carder, K., Arnone, R., He, M.-X., 2007. Determination of primary spectral bands for remote sensing of aquatic environments. Sensors 7 (12), 3428–3441.

160.Letelier, R.M., Abbott, M.R., 1996. An analysis of chlorophyll fluorescence algorithms for the moderate resolution imaging spectrometer (MODIS). Remote Sensing of Environment 58, 215–223.

161.Li, H. et al., 2010a. Enhancing generic ecological model for short-term prediction of Southern North Sea algal dynamics with remote sensing images. Ecological Modelling 221, 2435–2446.

162.Li, K. et al., 2010b. Spatial–temporal variation of chlorophyll-a concentration in the Bohai Sea. Lecture Notes in Computer Science 6330, 662–670.

163.Longhurst, A., 1995. Seasonal cycles of pelagic production and consumption. Progress in Oceanography 36, 77–167.

164.Lou, X., Hu, C., 2014. Diurnal changes of a harmful algal bloom in the East China Sea: observations from GOCI. Remote Sensing of Environment 140, 562–572.

165.Malone, T.C., 2008. Ecosystem dynamics, harmful algal blooms and operational oceanography. In: Babin, M. et al. (Eds.), Real-Time Coastal Observing Systems for Marine Ecosystem Dynamics and Harmful Algal Blooms: Theory, Instrumentation and Modelling. Oceanographic Methodology Series, UNESCO, Paris, pp. 527–560.

166.Marinelli, M.A. et al., 2008. A remote sensing study of the phytoplankton spatial– temporal cycle in the Southeastern Indian Ocean. Journal of Applied Remote Sensing, 2.

167.Maritorena, S. et al., 2002. Optimization of a semianalytical ocean color model for global applications. Applied Optics 41, 2705–2714.

168.Maritorena, S. et al., 2010. Merged satellite ocean color data products using a biooptical model: characteristics, benefits and issues. Remote Sensing of Environment 114, 1791–1804.

169.Marrari, M. et al., 2008. Spatial and temporal variability of SeaWiFS chlorophyll a distributions west of the Antarctic Peninsula: implications for krill production. Deep-Sea Research Part II 55, 377–392.

170.Martin, A.P., 2003. Phytoplankton patchiness: the role of lateral stirring and mixing. Progress in Oceanography 57, 125–174.

171.Martinez, E. et al., 2009. Climate-driven basin-scale decadal oscillations of oceanic phytoplankton. Science 326, 1253–1256.

172. Matthews, M.W., 2011. A current review of empirical procedures of remote sensing in Inland and near-coastal transitional waters. International Journal of Remote Sensing 32, 6855–6899.

173. Matthews, M.W. et al., 2012. An algorithm for detecting trophic status (chlorophylla), cyanobacterial-dominance, surface scums and floating vegetation in Inland and coastal waters. Remote Sensing of Environment 124, 637–652.

174. McClain, C.R., 2009a. A decade of satellite ocean color observations. Annual Review of Marine Science 1, 19–42.

175. McClain, C.R., 2009b. Satellite remote sensing: ocean color. In: Steele, J.H. et al. (Eds.), Encyclopedia of Ocean Sciences, second ed. Academic Press, Oxford, pp. 114–126.

176. McKee, D. et al., 2007. Potential Impacts of non-algal materials on water-leaving sun-induced chlorophyll fluorescence signals in coastal waters. Applied Optics 46, 7720–7729.

177. McKinna, L.I.W. et al., 2011. A simple, binary classification algorithm for the detection of Trichodesmium spp. within the great barrier reef using MODIS imagery. Limnology and Oceanography-Methods 9, 50–66.

178. Mélin, F. et al., 2011. Multi-sensor satellite time series of optical properties and chlorophyll-a concentration in the Adriatic Sea. Progress in Oceanography 91, 229–244.

179. Merico, A. et al., 2003. Analysis of satellite imagery for Emiliania huxleyi blooms in the Bering Sea before 1997. Geophysical Research Letters 30, 1337.

180. Miller, P.I. et al., 2006. SeaWiFS discrimination of harmful algal bloom evolution. International Journal of Remote Sensing 27, 2287–2301.

181. Moore, T.S. et al., 2009. A class-based approach to characterizing and mapping the uncertainty of the MODIS ocean chlorophyll product. Remote Sensing of Environment 113, 2424–2430.

182. Moore, T.S. et al., 2012. Detection of coccolithophore blooms in ocean color satellite imagery: a generalized approach for use with multiple sensors. Remote Sensing of Environment 117, 249–263.

183. Morel, A., 1980. In-water and remote measurements of ocean color. Boundary Layer Meteorology 18, 177–201.

184. Morel, A., 2001. Bio-optical models. In: Steele, J.H. et al. (Eds.), Encyclopedia of Ocean Sciences. Academic Press, New York, pp. 317–326.

185. Morel, A., Prieur, L., 1977. Analysis of variations in ocean color. Limnology and Oceanography 22, 709–722.

186. Moses, W.J. et al., 2012. Operational MERIS-based NIR-red algorithms for estimating chlorophyll-a concentrations in coastal waters—the Azov Sea case study. Remote Sensing of Environment 121, 118–124.

187.Mouw, C.B. et al., 2012. Impact of phytoplankton community size on a linked global ocean optical and ecosystem model. Journal of Marine Systems 89, 61– 75.

188.Navarro, G., Ruiz, J., 2006. Spatial and temporal variability of phytoplankton in the Gulf of Cádiz through remote sensing images. Deep Sea Research Part II: Topical Studies in Oceanography 53, 1241–1260.

189.Neville, R.A., Gower, J., 1977. Passive remote sensing of phytoplankton via chlorophyll fluorescence. Journal of Geophysical Research 82, 3487– 3493.

190.Nezlin, N.P., Li, B.-L., 2003. Time-series analysis of remote-sensed chlorophyll and cnvironmental factors in the Santo Monica—San Pedro Basin off Southern California. Journal of Marine Systems 39, 185–202.

191.Nezlin, N.P. et al., 1999. Patterns of seasonal and interannual changes of surface chlorophyll concentration in the black sea revealed from the remote sensed data. Remote Sensing of Environment 69, 43–55.

192.Nezlin, N.P. et al., 2010. Satellite monitoring of climatic factors regulating phytoplankton variability in the Arabian (Persian) Gulf. Journal of Marine Systems 82, 47–60.

193.Odermatt, D. et al., 2012. Review of constituent retrieval in optically deep and complex waters from satellite imagery. Remote Sensing of Environment 118, 116–126.

194.Oliveira, P.B. et al., 2009. Summer diatom and dinoflagellate blooms in Lisbon Bay from 2002 to 2005: pre-conditions inferred from wind and satellite data. Progress in Oceanography 83, 270–277.

195.Oliver, M.J., Irwin, A.J., 2008. Objective global ocean biogeographic provinces. Geophysical Research Letters, 35.

196.O'Reilly, J.E. et al., 1998. Ocean color chlorophyll algorithms for SeaWiFS. Journal of Geophysical Research 103, 24,937–24,954.

197.O'Reilly, J.E. et al., 2000. Ocean color chlorophyll-a algorithms for SeaWiFS, Oc2 and Oc4: version 4. In: Hooker, S.B., Firestone, E.R. (Eds.), SeaWiFS Postlaunch Technical Report Series. NASA Goddard Space Flight Center, Greenbelt (MD, USA), pp. 9–23.

198.Otero, M.P., Siegel, D., 2004. Spatial and temporal characteristics of sediment plumes and phytoplankton blooms in the Santa Barbara channel. Deep-Sea Research II 51, 1129–1149.

199.Pan, X.J. et al., 2012. Remote sensing of phytoplankton community composition along the northeast coast of the United States. Remote Sensing of Environment 115, 3731–3747.

200.Pan, X. et al., 2013. Remote sensing of picophytoplankton distribution in the Northern South China Sea. Remote Sensing of Environment 128, 162–175.

201.Park, J. et al., 2010a. Variability of SeaWiFS chlorophyll-a in the southwest Atlantic sector of the southern ocean: strong topographic effects and weak seasonality. Deep Sea Research Part I: Oceanographic Research Papers 57, 604–620.

202.Park, Y.J. et al., 2010b. Detection of algal blooms in European waters based on satellite chlorophyll data from MERIS and MODIS. International Journal of Remote Sensing 31, 6567–6583.

203.Peñaflor, E.L. et al., 2007. Detection of monsoonal phytoplankton blooms in Luzon Strait with MODIS data. Remote Sensing of Environment 109, 443–450.

204.Platt, T., Sathyendranath, S., 1999. Spatial structure of pelagic ecosystem processes in the global ocean. Ecosystems 2, 384–394.

205.Platt, T., Sathyendranath, S., 2008. Ecological indicators for the pelagic zone of the ocean from remote sensing. Remote Sensing of Environment.

206.Platt, T. et al., 2008. Operational estimation of primary production at large geographical scales. Remote Sensing of Environment 112, 3437–3448.

207.Platt, T. et al., 2009. The phenology of phytoplankton blooms: ecosystem indicators from remote sensing. Ecological Modelling 220, 3057–3069.

208.Qin, Y. et al., 2007. Validity of SeaDAS water constituents retrieval algorithms in Australian tropical coastal waters. Geophysical Research Letters, 34.

209.Quartly, G.D., Srokosz, M.A., 2004. Eddies in the southern Mozambique Channel. Deep Sea Research Part II: Topical Studies in Oceanography 51 (1–3), 69– 83.

210.Racault, M.-F. et al., 2012. Phytoplankton phenology in the global ocean. Ecological Indicators 14, 152–163.

211.Radenac, M.-H. et al., 2013. A very oligotrophic zone observed from space in the equatorial Pacific warm pool. Remote Sensing of Environment 134, 224–233.

212.Raitsos, D.E. et al., 2011. Assessing chlorophyll variability in relation to the environmental regime in Pagasitikos Gulf, Greece. Journal of Marine Systems 94, 16–22.

213.Raj, R.P. et al., 2010. Oceanic and atmospheric influences on the variability of phytoplankton bloom in the Southwestern Indian Ocean. Journal of Marine Systems 82, 217–229.

214.Richardson, L.L., 1996. Remote sensing of algal bloom dynamics. Bioscience 46, 492–501.

215.Richardson, K., 1997. Harmful or exceptional phytoplankton blooms in the marine ecosystem. Advances in Marine Biology 31, 301–385.

216.Richardson, A.J., Schoemann, D.S., 2004. Climate impact on plankton ecosystems in the Northeast Atlantic. Science 305, 1609–1612.

217. Robinson, I.S., 2004. Measuring the Oceans from Space: The Principle and Methods of Satellite Oceanography. Springer – Praxis, Chichester, UK. Robinson, C.L.K. et al., 2004. Twenty years of satellite observations describing phytoplankton blooms in seas adjacent to GwaiiHaanas National Park Reserve, Canada. Canadian Journal of Remote Sensing 30, 36–43.
218. Ruddick, K.G. et al., 2001. Optical remote sensing of chlorophyll a in case 2 waters by use of an adaptive two-band algorithm with optimal error properties. Applied Optics 40, 3575–3585.
219. Ruddick, K. et al., 2008. Overview of ocean colour: theoretical background, sensors and applicability to detection and monitoring of harmful algal blooms (capabilities and limitations). In: Babin, M. et al. (Eds.), Real-Time Coastal Observing Systems for Marine Ecosystem Dynamics and Harmful Algal Blooms. UNESCO Publishing, pp. 331–383.
220. Ryan, J.P. et al., 2008. A coastal ocean extreme bloom incubator. Geophysical Research Letters 35, L12602.
221. Ryan, J.P. et al., 2009. Influences of upwelling and downwelling winds on red tide bloom dynamics in Monterey Bay, California. Continental Shelf Research 29, 785–795.
222. Saitoh, S.-I. et al., 2002. A description of temporal and spatial variability in the Bering Sea spring phytoplankton blooms (1997–1999) using satellite multisensory remote sensing. Progress in Oceanography 55, 131–146.
223. Santoleri, R. et al., 2003. Year-to-year variability of the phytoplankton bloom in the Southern Adriatic Sea (1998–2000): sea-viewing wide field-of-view sensor observations and modeling study. Journal of Geophysical Research 108, 8122.
224. Sasamal, S.K. et al., 2005. Asterionella blooms in the Northwestern Bay of Bengal during 2004. International Journal of Remote Sensing 26, 3853–3858.
225. Sathyendranath, S. et al., 2001. Remote sensing of phytoplankton pigments: a comparison of empirical and theoretical approach. International Journal of Remote Sensing 22, 249–273.
226. Sathyendranath, S. et al., 2004. Discrimination of diatoms from other phytoplankton using ocean color data. Marine Ecology Progress Series 272, 59–68.
227. Scargle, J.D., 1982. Studies in astronomical time series analysis. Ii. Statistical aspects of unevenly spaced data. Astrophysical Journal, 757–763.
228. Sen Gupta, A., McNeil, B., 2012. Variability and change in the ocean. In: endersonSellers,A., McGuffie, K. (Eds.), The Future of the World's Climate, second ed. Elsevier, Boston, pp. 141–165.
229. Shang, S. et al., 2010. MODIS observed phytoplankton dynamics in the Taiwan Strait: an absorption-based analysis. Biogeosciences Discussions 7, 7795–7819.

230. Shanmugam, P., 2011. A new bio-optical algorithm for the remote sensing of algal blooms in complex ocean waters. Journal of Geophysical Research-Oceans 116, 12.
231. Shanmugam, P. et al., 2008. SeaWiFS sensing of hazardous algal blooms and their underlying mechanisms in shelf-slope waters of the Northwest Pacific during summer. Remote Sensing of Environment 112, 3248–3270.
232. Shen, L. et al., 2012. Satellite remote sensing of harmful algal blooms (HAB) and a potential synthesized framework. Sensors 12, 7778–7803.
233. Shevyrnogov, A. et al., 2002a. Spatial and temporal anomalies of chlorophyll concentration in Atlantic Ocean (by space-based data). Advances in Space Research 30, 2541–2546.
234. Shevyrnogov, A. et al., 2002b. Trends of chlorophyll concentration in the surface layer of the northern and central Atlantic, a satellite data-based study. Advances in Space Research 30, 2535–2540.
235. Shi, W., Wang, M., 2007. Observations of a Hurricane Katrina-induced phytoplankton bloom in the Gulf of Mexico. Geophysical Research Letters, 34.
236. Shutler, J.D. et al., 2010. Coccolithophore bloom detection in the north east Atlantic using SeaWiFS: algorithm description, application and sensitivity analysis. Remote Sensing of Environment 114, 1008–1016.
237. Shutler, J.D. et al., 2012. An adaptive approach to detect high-biomass algal blooms from EO chlorophyll-a data in support of harmful algal bloom monitoring. Remote Sensing Letters 3, 101–110.
238. Siegel, H. et al., 1999. Case studies on phytoplankton blooms in coastal and open waters of the Baltic Sea using coastal zone color scanner data. International Journal of Remote Sensing 20, 1249–1264.
239. Siegel, D.A. et al., 2002. The North Atlantic spring phytoplankton bloom and Sverdrup's critical depth hypothesis. Science 296, 730–733.
240. Siegel, H. et al., 2007. Identification of coccolithophore blooms in the SE Atlantic Ocean off Namibia by satellites and in-situ methods. Continental Shelf Research 27, 258–274.
241. Siegel, D.A. et al., 2013. Regional to Global assessments of phytoplankton dynamics from the SeaWiFS mission. Remote Sensing of Environment 135, 77–91.
242. Signorini, S.R., McClain, C.R., 2009. Environmental factors controlling the Barents Sea spring–summer phytoplankton blooms. Geophysical Research Letters 36, L10604.
243. Siswanto, E. et al., 2013. Detection of harmful algal blooms of Karenia Mikimotoi using MODIS measurements: a case study of Seto-Inland Sea, Japan. Remote Sensing of Environment 129, 185–196.

244. Smetacek, V., Cloern, J.E., 2008. On phytoplankton trends: how are phytoplankton at coastal sites around the world responding to ongoing global change? Science 319, 1346–1348.

245. Song, H. et al., 2010. Phenology of phytoplankton blooms in the Nova Scotian shelfGulf of Maine region: remote sensing and modeling analysis. Journal of Plankton Research 32, 1485–1499.

246. Sournia, A. et al., 1991. Marine phytoplankton: how many species in the world ocean? Journal of Plankton Research 13, 1093–1099.

247. Srokosz, M.A., Quartly, G.D., 2013. The Madagascar Bloom – a serendipitous study. Journal of Geophysical Research 118 (1), 14–25.

248. Srokosz, M.A., Quartly, G.D., Buck, J.J.H., 2004. A possible plankton wave in the Indian Ocean. Geophysical Research Letters 31 (13), L13301.

249. SteemannNielsen, E., 1937. The Annual Amount of Organic Matter Produced by the Phytoplankton in the Sound off Helsingor. MeddelelserfraKommissionen for DanmarksFiskeriogHavundersogelser. Ser. Plankton, pp. 1–37.

250. Steemann Nielsen, E., 1963. Productivity, definition and measurement. In: Hill, M.N. (Ed.), The Sea. Wiley, London.

251. Steemann Nielsen, E., Jensen, E., 1957. Primary Ocean Production. The Autotrophic Production of Organic Matter in the Oceans. Galathea Report, pp. 49–136.

252. Stumpf, R.P., 2001. Applications of satellite ocean color sensors for monitoring and predicting harmful algal blooms. Human and Ecological Risk Assessment 7, 1363–1368.

253. Stumpf, R.P. et al., 2003. Monitoring Karenia brevis blooms in the Gulf of Mexico using satellite ocean color imagery and other data. Harmful Algae 2, 147–160.

254. Subramaniam, A., Carpenter, E.J., 1994. An empirically derived protocol for the detection of blooms of the marine cyanobacteria Trichodesmium using CZCS imagery. International Journal of Remote Sensing 15, 1559–1569.

255. Subramaniam, A. et al., 1999a. Bio-optical properties of the marine diazotrophic cyanobacteria Trichodesmium spp.; Ii. A reflectance model for remote-sensing. Limnology and Oceanography 44, 618–627.

256. Subramaniam, A. et al., 1999b. Bio-optical properties of the marine diazotrophic cyanobacteria Trichodesmium spp.; I. Absorption and photosynthetic action spectra. Limnology and Oceanography 44, 608–617.

257. Subramaniam, A. et al., 2002. Detecting Trichodesmium blooms in SeaWiFS imagery. Deep-Sea Research II 49, 107–121.

258. Tan, C.K. et al., 2006. Seasonal variability of SeaWiFS chlorophyll a in the Malacca Straits in relation to Asian monsoon. Continental Shelf Research 26, 168–178.

259. Tang, D.-L. et al., 1999. Remote sensing observations of winter phytoplankton blooms southwest of the Luzon Strait in the South China Sea. Marine Ecology Progress Series 191, 45–51.
260. Tang, D.L. et al., 2004. Long-time observation of annual variation of Taiwan Strait upwelling in summer season. Advances in Space Research 33, 307–312.
261. Tang, D. et al., 2006. Satellite evidence of harmful algal blooms and related oceanographic features in the Bohai Sea during Autumn 1998. Advances in Space Research 37, 681–689.
262. Thomas, A.C. et al., 2003. Satellite-measured phytoplankton variability in the Gulf of Maine. Continental Shelf Research 23, 971–989.
263. Thomas, A.C. et al., 2012. Satellite views of Pacific chlorophyll variability: comparisons to physical variability, local versus nonlocal influences and links to climate indices. Deep Sea Research Part II: Topical Studies in Oceanography, 10.
264. Tilstone, G. et al., 2011a. Variability in specific-absorption properties and their use in a semi-analytical ocean colour algorithm for MERIS in North Sea and Western English Channel coastal waters. Remote Sensing of Environment 118, 320–338.
265. Tilstone, G.H. et al., 2011b. An assessment of chlorophyll-a algorithms available for SeaWiFS in coastal and open areas of the Bay of Bengal and Arabian Sea. Remote Sensing of Environment.
266. Tiwari, S.P., Shanmugam, P., 2013. Evaluation of inversion models for the satellite retrieval of absorption coefficient of phytoplankton in oceanic/coastal waters. IEEE Journal of Selected Topics in Applied Earth Observations and Remote Sensing, 99.
267. Tomlinson, M.C. et al., 2004. Evaluation of the use of SeaWiFS imagery for detecting Karenia brevis harmful algal blooms in the Eastern Gulf of Mexico. Remote Sensing of Environment 91, 293–303.
268. Tomlinson, M.C. et al., 2008. An evaluation of remote sensing techniques for enhanced detection of the toxic dinoflagellate, Karenia brevis. Remote Sensing of Environment 113, 598–609.
269. Torrecilla, E. et al., 2011. Cluster analysis of hyperspectral optical data for discriminating phytoplankton pigment assemblages in the open ocean. Remote Sensing of Environment 115, 2578–2593.
270. Tyrrell, T. et al., 2005. Effect of seafloor depth on phytoplankton blooms in highnitrate, low-chlorophyll (HNLC) regions. Journal of Geophysical Research, 110.
271. Uiboupin, R. et al., 2012. Monitoring the effect of upwelling on the chlorophyll a distribution in the Gulf of Finland (Baltic Sea) using remote sensing and in situ data. Oceanologia 54, 395–419.

272. Uitz, J. et al., 2006. From surface chlorophyll a to phytoplankton community composition in oceanic waters. Journal of Geophysical Research, 111.

273. Urquhart, E.A. et al., 2013. Geospatial interpolation of MODIS-derived salinity and temperature in the Chesapeake Bay. Remote Sensing of Environment 135, 167– 177.

274. Uz, B.M., 2007. What causes the sporadic phytoplankton bloom Southeast of Madagascar? Journal of Geophysical Research (Oceans) 112, C09010.

275. Vargas, M. et al., 2009. Phenology of marine phytoplankton from satellite ocean color measurements. Geophysical Research Letters, 36.

276. Venables, H.J. et al., 2007. Physical conditions controlling the development of a regular phytoplankton bloom North of the Crozet Plateau, Southern Ocean. Deep Sea Research Part II: Topical Studies in Oceanography 54, 1949–1965.

277. Villareal, T.A., Adornato, L., Wilson, C., Schoenbaechler, C.A., 2011. Summer blooms of diatom-diazotroph assemblages and surface chlorophyll in the North Pacific gyre: a disconnect. Journal of Geophysical Research 116 (C3), C03001.

278. Villareal, T.A. et al., 2012. Summer diatom blooms in the North Pacific Subtropical Gyre: 2008–2009. PLoS One 7, e33109.

279. Vinayachandran, P.N., Mathew, S., 2003. Phytoplankton bloom in the Bay of Bengal during the northeast monsoon and its intensification by cyclones. Geophysical Research Letters 30, 1572.

280. Volpe, G. et al., 2007. The colour of the Mediterranean Sea: global versus regional bio-optical algorithms evaluation and implication for satellite chlorophyll estimates. Remote Sensing of Environment 107, 625–638.

281. Voss, K.J. et al., 1998. Scattering and attenuation properties of Emiliania huxleyi cells and their detached coccoliths. Limnology and Oceanography 43, 870–876.

282. Wan, Z. et al., 2013. Comparison of two light attenuation parameterization focusing on timing of spring bloom and primary production in the Baltic Sea. Ecological Modelling 259, 40–49.

283. Wang, D., Zhao, H., 2008. Estimation of phytoplankton responses to Hurricane Gonu over the Arabian Sea based on ocean color data. Sensors 8, 4878–4893.

284. Wang, J. et al., 2010. Winter phytoplankton bloom induced by subsurface upwelling and mixed layer entrainment southwest of Luzon Strait. Journal of Marine Systems 83, 141–149.

285. Westberry, T.K. et al., 2005. An improved bio-optical model for the remote sensing of Trichodesmium spp. blooms. Journal of Geophysical Research 110 (C06012), 1–11.

286. Westberry, T.K. et al., 2013. Retrospective satellite ocean color analysis of purposeful and natural ocean iron fertilization. Deep Sea Research Part I: Oceanographic Research Papers 73, 1–16.
287. White, A.E. et al., 2007. What factors are driving summer phytoplankton blooms in the North Pacific Subtropical Gyre? Journal of Geophysical Research (Oceans), 112.
288. Wilson, C., 2003. Late summer chlorophyll blooms in the oligotrophic North Pacific Subtropical Gyre. Geophysical Research Letters 30 (18).
289. Wilson, C., 2011. The rocky road from research to operations for satellite oceancolour data in fishery management. ICES Journal of Marine Science 68, 677– 686.
290. Wilson, C., Qiu, X., 2008. Global distribution of summer chlorophyll blooms in the oligotrophic gyres. Progress in Oceanography 78 (2), 107–134.
291. Wilson, C., Villareal, T.A., Maximenko, N., Bograd, S.J., Montoya, J.P., Schoenbaechler, C.A., 2008. Biological and physical forcings of late summer chlorophyll blooms at 30N in the oligotrophic Pacific. Journal of Marine Systems 69 (3–4), 164–176.
292. Wong, K.T.M. et al., 2009. Forecasting of environmental risk maps of coastal algal blooms. Harmful Algae 8, 407–420.
293. Xing, X.-G. et al., 2007. An overview of remote sensing of chlorophyll fluorescence. Ocean Science Journal 42, 49–59.
294. Yentsch, C.S., Menzel, D.W., 1963. A method for the determination of phytoplankton chlorophyll and phaeophytin by fluorescence. Deep Sea Research and Oceanographic Abstracts 10, 221–231.
295. Yoder, J.A. et al., 2001. Variability in coastal zone color scanner (Czcs) chlorophyll imagery of ocean margin waters off the US east coast. Continental Shelf Research 21, 1191–1218.
296. Yoo, S. et al., 2008. Seasonal, interannual and event scale variation in North Pacific ecosystems. Progress in Oceanography 77, 155–181.
297. Yuan, J. et al., 2005. In-pixel variations of chl a fluorescence in the Northern Gulf of Mexico and their implications for calibrating remotely sensed chl a and other products. Continental Shelf Research 25, 1894–1904.
298. Zeichen, M.M., Robinson, I.S., 2004. Detection and monitoring of algal blooms using SeaWiFS imagery. International Journal of Remote Sensing 25, 1389–1395.
299. Zhao, H. et al., 2008. Comparison of phytoplankton blooms triggered by two typhoons with different intensities and translation speeds in the South China Sea. Marine Ecology Progress Series 365, 57–65.
300. Zhao, H. et al., 2009a. Phytoplankton Blooms near the Pearl River estuary induced by Typhoon Nuri. Journal of Geophysical Research 114 (12).

301.Zhao, J. et al., 2009b. The variations in optical properties of CDOM throughout an algal bloom event. Estuarine, Coastal and Shelf Science 82, 225–232.

302.Zhao, D. et al., 2010. The relation of chlorophyll-a concentration with the reflectance peak near 700 nm in algae-dominated waters and sensitivity of fluorescence algorithms for detecting algal bloom. International Journal of Remote Sensing 31 (10).

303.Zhong, Y., Bracco, A., Villareal, T.A., 2012. Pattern formation at the ocean surface: Sargassum distribution and the role of the eddy field. Limnology and Oceanography: Fluids & Environments 2, 12–27. http://dx.doi.org/10.1215/21573689-1573372.

304.Zingone, A. et al., 2010. Multiscale variability of twenty-two coastal phytoplankton time series: a global scale comparison. Estuaries and Coasts 33, 224–229.

CITATION

David Blondeau-Patissier, James F.R. Gower, Arnold G. Dekker, Stuart R. Phinn, Vittorio E. Brando, A review of ocean color remote sensing methods and statistical techniques for the detection, mapping and analysis of phytoplankton blooms in coastal and open oceans, Progress in Oceanography, Volume 123, April 2014, Pages 123-144, ISSN 0079-6611, http://dx.doi.org/10.1016/j.pocean.2013.12.008.

CHAPTER 3

Land Resources Assessment of El-Galaba Basin, South Egypt for the Potentiality of Agriculture Expansion Using Remote Sensing and Gis Techniques

A.M. Saleh, A.B. Belal, E.S. Mohamed

Soil Sciences Dept., National Authority for Remote Sensing and Space Sciences (NARSS), Cairo, Egypt

ABSTRACT

The socio-economic development in Egypt is based on land resources. Recently, the Egyptian government is interested in developing low desert zone areas which are located between the recent Nile flood plain and the limestone plateau, from the east and west sides, and represent an important source of aggregate materials. Therefore, this study was carried out to investigate the potentiality of El-Galaba basin soils which are located in the western part of the Aswan Governorate and are characterized by Wadi El-Kubbaniya for the horizontal agricultural expansion and their optimum agricultural use. The investigated area was remotely sensed to identify the landscape and its land resources. Terrain units were identified using draped Landsat 8 satellite image over Digital Terrain Model (DTM) to express the landscape and the associated soil mapping units. Fifteen mapping units were identified and grouped. Land capability evaluation was performed using Cervatana capability model. The results of capability modeling revealed about 3.33% of land with good use capability, 76.06% land with moderate use capability, and 0.08% marginal or non-productive land. The main capability limitations were soil and erosion risks. The Almagra model was used to produce the optimum cropping pattern and limitations of soil units. Matching the crop requirements with soil characteristics, optimum cropping pattern was obtained for wheat, corn, melon, potatoes, sunflower, sugar beet, Alfalfa, peach, citrus, and olive. The results of the study revealed the potentiality of El-Galaba basin for agricultural uses.

INTRODUCTION

Egypt is a net food importer, including for over half of its wheat needs. The increasing population and limited cultivated land, combined with land degradation and desertification pose significant challenges for production. Between 2010 and 2011 the total cultivated area in Egypt decreased by 1 percent, associated with encroachment on agricultural land (World Food Programme, 2013). Upper Egypt experienced more severe and unpredictable weather and crop failures, leading farmers to over-utilize already stressed natural resources to compensate for low productivity. Egypt has seen an increase in food insecurity as the prevalence of combined food insecurity and income poverty in Egypt raised to 17.2% (an estimated 13.7 million people) in 2011, up from 14 percent of the population in 2009. The rainfall in the Western Desert of Egypt (Sahara) is very rare and occurs mainly from cyclonic winter storms, which could occur once every 10–20 years (Goudie, 2002 and Brookes, 2003). The estimates predicted economic losses as a result of climatic change as up to 14% of the African GDP if adaptation measures fail to be implemented (Clements, 2009 and Nelson et al., 2009). Agriculture was the most affected sector as a result of climatic variability. The required increase in agricultural production to meet future food demand will further increase pressure on land resources.

Remote sensing techniques have been utilized in soil science for many years as a tool for soil surveyors, reducing the time and expense for sampling (Palacios-Orueta and Ustin, 1998). Geographic information system (GIS) plays a major role in spatial decision-making. The collected information for the suitability analysis for crop production should present both opportunities and constraints for the decision maker (Ghafari et al., 2000). The ultimate aim of GIS is to provide support for spatial decisions making process (Foote and Lynch, 1996). Data layers in multi-criteria evaluation are handled in order to arrive at the suitability, which can be conveniently achieved using GIS. Remote sensing and GIS were used in many studies in Egypt for land resources mapping and management (Saleh et al., 2013, Mohamed et al., 2013, Mohamed et al., 2014 and Saleh and Belal, 2014) .The process of land suitability classification is the evaluation and grouping of specific areas of land in terms of their suitability for a defined land use. The suitability defines the level of crop requirements with respect to the present soil characteristics. The suitability is a measure of how well the qualities of a land unit match with the requirements of a particular form of land use (FAO, 2003). Interpreting soil qualities and site information for the agriculture use and management practices is integrated using GIS (FAO, 1991 and FAO, 2007). The land quality is a complex attribute of land which acts in a manner distinct from the actions of other land qualities in its influence on the land suitability for a specific kind of use (FAO, 1985); it is the ability of land to fulfill specific requirements for the land utilization type (LUT) (Van Diepen et al., 1991). Spatial analysis can be defined as the analytical techniques associated with the study of geographic phenomena locations together with their spatial dimensions and their associated attributes (ESRI, 2001).

This study was carried out to investigate the potentiality of El-Galaba basin soils, which are a part of the Western Desert of Egypt and located in the western part of the Aswan Governorate, for the horizontal agricultural expansion and their optimum agricultural use.

MATERIAL AND METHODS

Study area

The study area is located west of Aswan, in the Western Desert and extends between latitude 24°10′ and 25°10′N and longitude 32°10′ to 33°E, and comprises the El-Galaba plain, the western part of the Kom Ombo basin, and Wadi El-Kubbaniya (Map 1). The study area of El-Galaba plain is a desert area with no landuse. The climatic conditions of El-Galaba area are typically a desert-like condition characterized by an extremely arid climate with long hot rainless summer, mild winter with – very low to no amount of rainfall. The mean temperature in winter ranges from 16.1 to 18.7 °C and 15.9 to 17.3 °C in Aswan and Kom Ombo stations, respectively. In summer, the mean temperature ranges from 32.9 °C to 33.4 °C and 32.4 to 33.0 °C in Aswan and Kom Ombo stations, respectively. The intensity of rainfall at El-Galaba area is very low to negligible as the annual intensity reaches 0.7 mm in Aswan and 1.2 mm in Kom Ombo stations, respectively, while no rain is recorded in summer. Although rainfall is not significant throughout the year, some rare and irregular storms take place over scattered localities during the winter season.

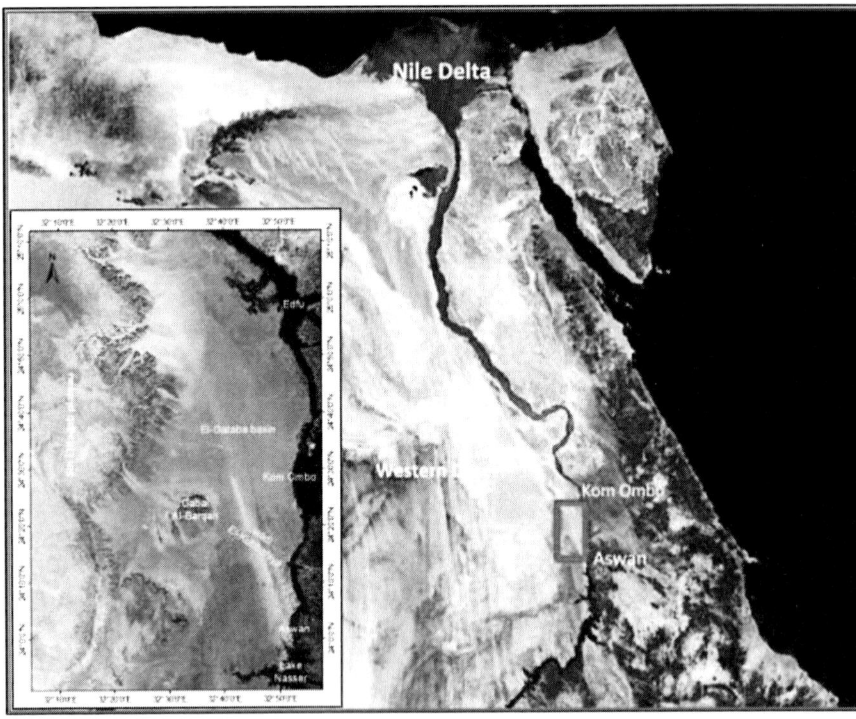

Map 1. Location map of El-Galaba plain.

From the geologic point of view, the Galaba area is situated within the African Platform with its Pre-Cambrian folded basement, the tectonic framework is related to the Last African Orogenic belt (Said, 1962 and Abd El-Razik and Razavaliaev, 1972). The Nile Valley in Egypt is controlled by wrench faults that are generally parallel either to the Gulf of Suez or the Gulf of Aqaba directions (Youssef, 1968). The western region of the Nile valley consists of a series of unconnected depressions (Attia, 1954 and Said, 1962. andFAO, 1966). The study area has been affected by the same structural deformation processes that generated the Nile Valley and shaped the Kom Ombo basin located east of the Nile River (Koch et al., 2012). The Nubia sandstone sequence in the study area consists of alternating beds of sandstones, shale and clay (Issawi, 1983 and Issawi and McCauly, 1993). The faults in the study area represent two main faulting sets; NNW-SSE and NW-SE and in addition there are some major E-W trending faults. These faults are mainly of gravity types and are responsible for the development of the horst and graben structures in the studied area (Salman, 1974).

Digital image processing

The following remote sensing analyses (Fig. 1) used data from Landsat Data Continuity Mission (LDCM) sensor (Landsat 8) covering the study area acquired under clear sky conditions dated to the year 2013. Data of the blue to the short-wave infrared portion of the spectrum were used in this study. At first, the Thermal InfraRed Sensor (TIRS) bands, the coastal and the cirrus bands were excluded due to the nature of the study. The 30-m spatial resolution of Operational Land Imager (OLI) data were resampled to the higher 15-m resolution of the panchromatic band. All further digital image processing and analyses were executed using the standard approaches provided by the software ENVI 5.0 (Exelis Visual Information Solutions, 2012). The processing of data was represented by rescaling the OLI data to the Top of Atmosphere (TOA) radiance using the algorithm provided by (USGS, 2013) and the radiometric rescaling coefficients provided in the product metadata file. The OLI data were classified using the ISO-DATA classification technique (Map 2). The ISO-DATA classifier was used as it calculates the class means evenly distributed in the data space then iteratively clusters the remaining pixels using minimum distance techniques (Lillesand and Kiefer, 2000). Post classification procedures were performed on the resulted classified image to enhance the classification result. These procedures conclude Sieving, Clumping, and Majority analyses.

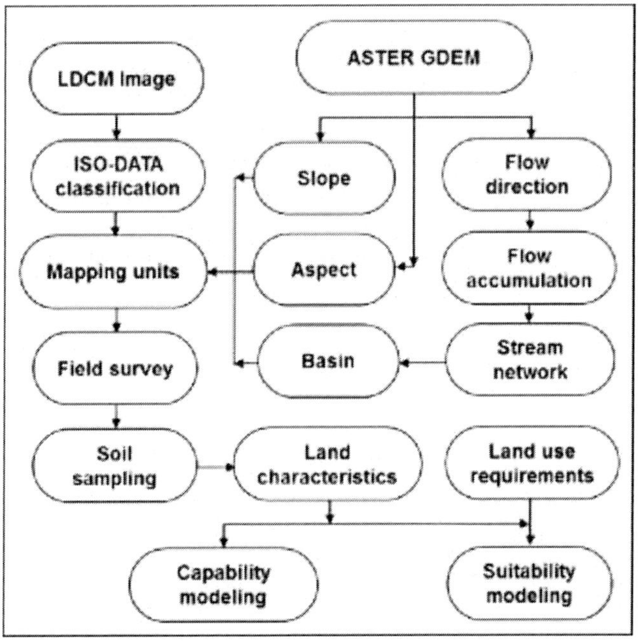

Figure 1. Framework of the methodology followed in the study.

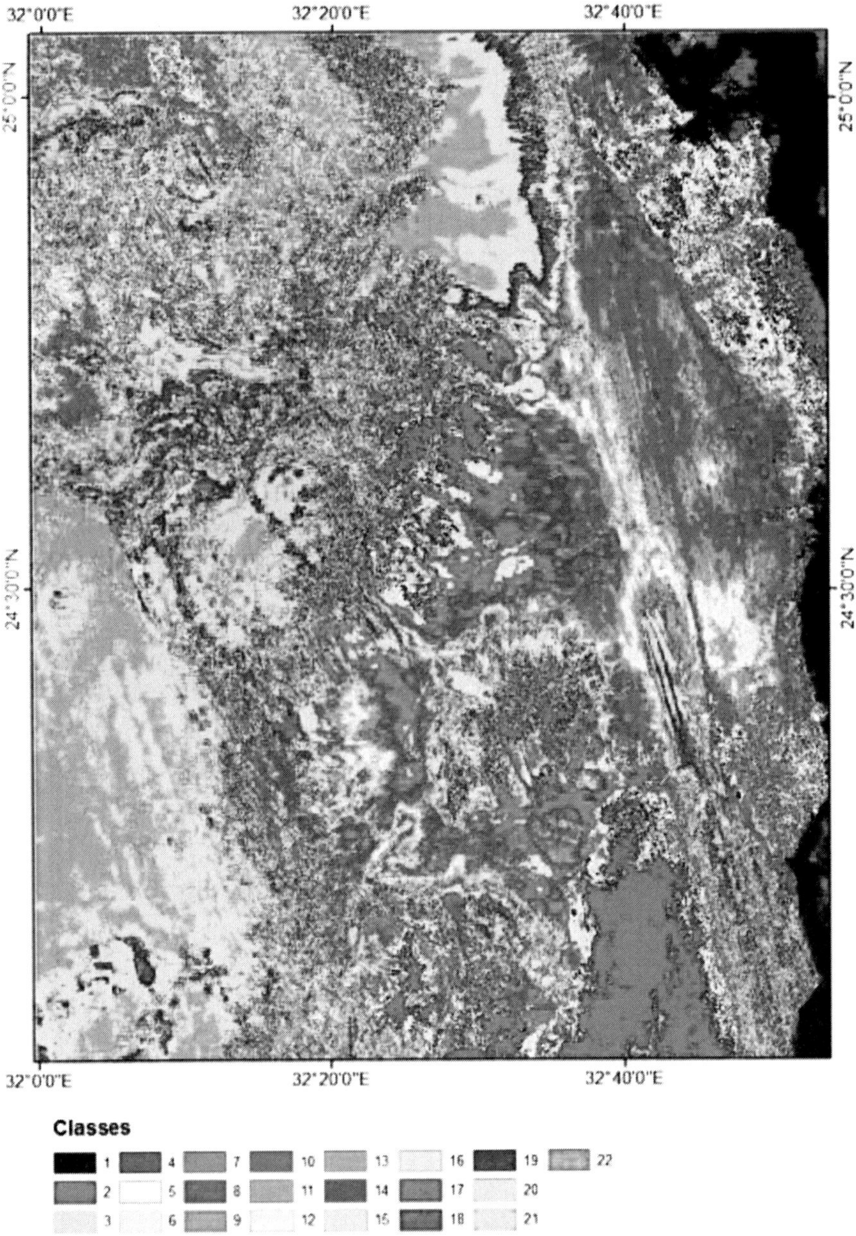

Map 2. Land classes of El-Galaba plain.

Digital terrain analyses

Digital Terrain Model (DTM) of El-Galaba study area was generated and prepared for the subsequent analyses (Fig. 1). Advanced Spaceborne Thermal Emission and Reflection Radiometer (ASTER) Global Digital Elevation Model ASTER (GDEM) v2 1 arc-second image was used as the source data for elevation heights of the study area. Digital elevation analyses were performed on ASTER GDEM data to extract parametric information, including slope, aspect, flow direction, flow accumulation, stream networks and watersheds (Wood, 1996) using ArcMap 10.2 (ESRI, 2013). Stream networks were derived by applying an area threshold value (100 pixels) to the output from the flow accumulation function using a grid algebraic expression (Tarboton et al., 1991). The stream network links were classified and the stream's order was extracted using the method proposed by (Strahler, 1957). The raster network converted to arc coverage. El-Galaba basin was derived by first identifying the outlet cells above which the basin will be determined, and then using the identified outlets and the flow direction grids to derive the basin utilizing algorithms developed by Jenson and Domingue (1988).

Mapping units extraction

The Digital Terrain Model (DTM) was analyzed with the aid of the satellite image analysis, the previous DTM extracted parametric information (slope, aspect, and stream networks), the existing body of knowledge in geomorphology, and field observations to obtain the mapping units of El-Galaba basin. The classified satellite image was draped over the Digital Terrain Model (DTM) to get a natural 3D theme to get a better understanding of the mapping units and to facilitate extracting these units (Fig. 1).

Based on the field survey, the digital terrain analyses and the soil analyses, the obtained land classes were combined into a number of fifteen mapping units. The resulted mapping units were identified and grouped. Thirteen mapping units were identified as soil mapping units (SMU) while the other two mapping units were identified as rocky non-soil mapping units (RMU). The Sin El-Kadab plateau which represents the western borders of El-Galaba basin was excluded from the mapping units. The mapping units were verified during the field survey. Each mapping unit of El-Galaba basin was identified on the map by a symbol. Consistent nomenclature is essential for understanding the relationships and differences among mapping units and for correlating the soil units with those found elsewhere, in order to make use of the whole body of existing knowledge about soil genesis and behaviors.

Field survey

A rapid reconnaissance survey was made throughout the investigated area in order to gain an appreciation of the broad soil patterns and characteristic landscape of El-Galaba basin. The primary mapping units were verified based on the pre-field interpretation and the information gained during the survey. Secondly, a number of thirty-three soil profiles were dug representing the different soil mapping units of El-Galaba basin. The soil profiles were dug to a depth of 150 cm, unless obstructed or hindered by bedrock. Soil samples representing the subsequent variations within the soil horizons were collected for laboratory analyses of some physical and chemical soil properties. The soil samples were thoroughly examined and morphologically described in the selected sites according to the system outlined by (FAO, 2006). The soil profiles were geo-referenced using GPS "MAGELLAN-GPS NAV DLX-10 TM".

Soil analyses

The soil samples were air-dried, crushed softly, and passed through a 2-mm sieve to get the "fine earth." The fine earth was analyzed in the laboratory for physical and chemical analyses, that included particle size distribution, percentage fine sand, silt, clay, organic matter, pH, electrical conductivity, calcium carbonates content, etc. carried out according to (Soil Survey Staff, 2014).

Land capability modeling

A land capability modeling procedure was applied following the generally accepted MicroLEIS capability model Cervatana (De la Rosa et al., 2004). The Cervatana model forecasts the general land use capability for a broad series of possible agricultural uses following the generally accepted norms of land evaluation (FAO, 1976, Dent and Young, 1981, ONERN, 1982 and Verheye, 1986). The methodological criteria refer to the system designed earlier by (De la Rosa and Magaldi, 1992) and modified for computing purposes by De la Rosa et al. (2004). The prediction of general land use capability is the result of a qualitative evaluation process or overall interpretation of the following biophysical factors: relief, soil, climate, and current use or vegetation (Fig. 2). For each diagnostic criterion or limiting factor, the land characteristics were selected, and the corresponding levels of generalization were established and related with the capability classes by means of gradation matrices (Table 1). The procedure of maximum limitation was used with matrices of degree to relate the land characteristics directly with capability classes. Matching tables were used to express inferences and define four capability classes:

Class S1 (Excellent); Class S2 (Good); Class S3 (Moderate); and Class N (Marginal). Four subclasses are also defined according to site (t), soil (l), erosion risks (r), and bioclimatic deficiency (b) limitations. Matching tables were used and linked to the ARCGIS modeling environment using relational database fields which have identifier key attribute property.

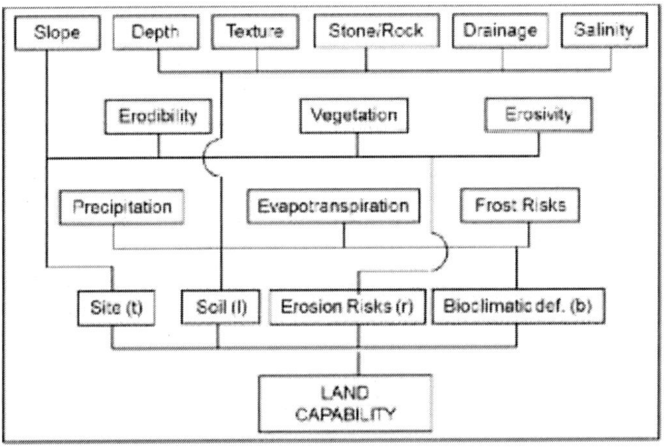

Figure 2. Flowchart of Cervatana for predicting land capability.

Table 1.Gradation matrix between the levels of generalization established for the limiting factors and classes of land use capability.

Limiting factor	Levels of generalization	Land use capability class			
		S1	S2	S3	N
Slope (t)	Nil or gentle	<7	–	–	–
%	Slight to moderate	–	7–15	–	–
	Strong	–	–	15–30	–
	Steep	–	–	–	>30
Soil	Useful depth (cm)	High (>75)	Moderate (50–75)	Shallow (25–50)	Superficial (<25)
	Texture	Balanced	Slight or heavy	–	–
	stoniness/rockiness (%)	Nil or slight (<15)	Slight or moderate (15–40)	High (>40)	–

	Drainage	Good	Moderate	Deficit or excessive	–
	salinity (dS/m)	Nil or slight (<4)	Moderate (4–8)	High (8–12)	Very high (>12)
Erosion risk (r)	Erodibility of soil	Slight	Moderate	High (>30)	–
	Slope	<15	15–30	–	–
	Density of vegetation	High	Moderate	Nil	–
	Erosivity	Slight (<150)	Moderate (150–200)	Strong (200–300)	Very strong (>300)
Bioclimatic deficiency (b)	Water deficiency	Low (Class h1)	Moderate (Class h2)	High (Class h3)	Very High (Class h4)
	Frost risk	Nil to slight (Class f1, f2)	Slight to moderate (Class f3)	High (Class f4)	–

Cropping pattern modeling

A land evaluation modeling for cropping pattern was applied following the MicroLEIS model Almagra (De la Rosa et al., 1992 and De la Rosa et al., 2004). The Almagra model (Fig. 3) indicates the degree of suitability for a land use irrespective of economic conditions. The land use requirements were matched to the land characteristics of each mapping unit to determine its suitability depending on the gradations considered for selected criteria (gradation matrices). The suitability classes for each crop are: soils with optimum suitability (S1), soils with high suitability (S2), soils with moderate suitability (S3), soils with marginal suitability (S4), and soils with no suitability (S5). The main soil limitations are: useful depth (p), texture (t), drainage condition (d), carbonates content (c), salinity (s), sodium saturation (a) and degree of development of the profile (g). For each diagnostic criterion or limiting factor, the land characteristics were selected, and the corresponding levels of generalization were established and related with the suitability classes by means of gradation matrices. The overall soil suitability of a soil component (unit) was assessed through the maximum limitation method where suitability is taken from the most limiting factor of soil characteristics. Ten land uses were tested for their suitability in the study area, namely, wheat, corn, melon, potatoes, sunflower, sugar beet, Alfalfa, peach, citrus, olive. The requirements of each kind of land use are obtained from (Sys et al., 1993).

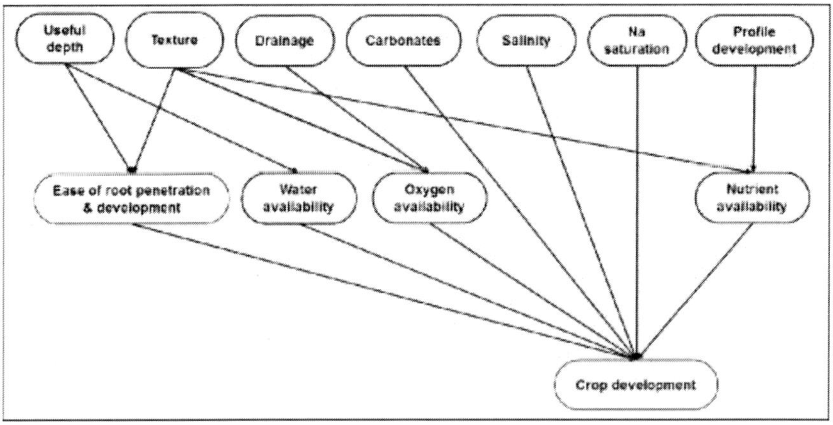

Figue 3. General scheme of the Almagra model, showing the direct and indirect effects of the selected characteristics on crop production via qualities.

RESULTS

Landscape characteristics

The landscape of El-Galaba basin is characterized generally by flat plain, ridges and small hills intersected by short valleys and wadis as well as the large Wadi El-Kubbaniya. This generally low relief seems to have formed in the Pleistocene (Issawi and Hinnawi, 1980). The digital elevation model analyses revealed that the ground surface elevation varies from about 600 m above sea level (asl) at Sin EL-Kadab plateau to about 77 m (asl) at the basin low areas (Map 3). The slope analyses revealed that, 37.75% of the area is flat to nearly level (0–1%), 55.16% of the area are very gently slope to sloping (1–10%), 6.60% of the area are strongly sloping to moderately steep (10–30%), and 0.49% of the area are steep (>30%). The aspect analyses resulted that, the directions of the slopes are flat (3.77%), north (5.42%), northeast (15.91%), east (17.00%), southeast (14.14%), south (11.48%), southwest (10.28%), west (10.88%), and northwest (11.12%). The highest stream order is the seventh order (Map 3). There is essentially no natural vegetation at the basin except for Wadi El-

Kubbaniya which has some shrubs in the middle of the drainage, but none of the short wadis show any vegetation.

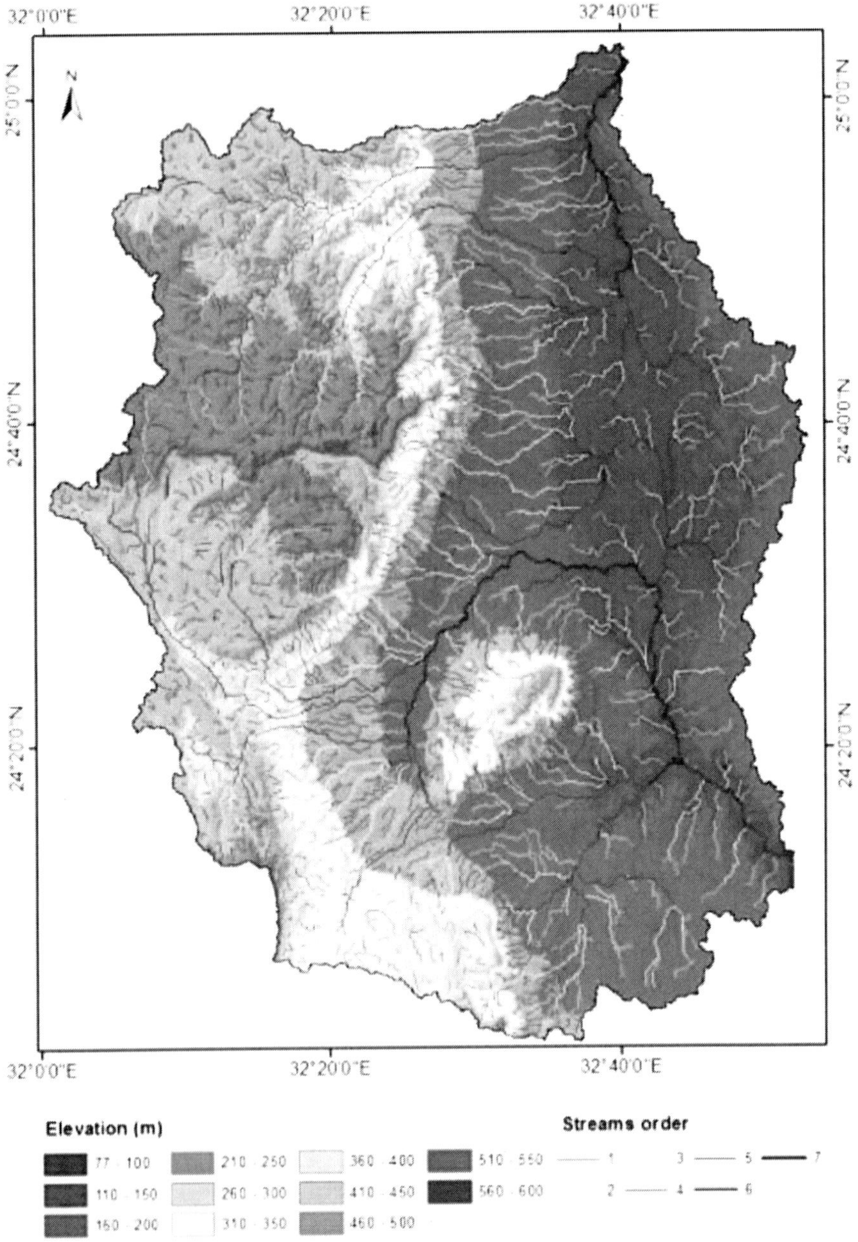

Map 3. Digital Terrain Model (DTM) overlaid with stream network of El-Galaba basin.

Mapping units of El-Galaba basin

Fifteen mapping units were extracted, identified and grouped in El-Galaba basin. Thirteen mapping units were identified as soil mapping units (SMU) while the other two mapping units were identified as rocky non-soil mapping units (RMU).

The soil mapping units (SMU) of El-Galaba basin and their associated soils were identified and delineated as shown in (Map 4). The soil mapping units (SMU) were grouped based on the soil horizons/layers similar in differentiating characteristics and arrangement in the soil profiles, except for the texture of the surface horizon, and developed from a particular type of parent material. They are equivalent to the soil series of the American soil taxonomy (Soil Survey Staff, 1999). The soils within the soil units are essentially homogeneous in all soil characteristics.

The following is a brief description of the mapping units of El-Galaba basin:

- Soil mapping unit 1 (SMU1): This mapping unit is deep, sandy, non-saline, with many gravels and some scattered boulders on the surface. It occupies an area of 223.19 km^2 (5.68%). The calcium carbonate content ranges between 1.63 and 2.71% in the different layers of the representative profiles. The organic matter content ranges between nil and 1.38%. The soil salinity values reveal that, the electrical conductivity (ECe) ranges between 0.33 and 0.82 dS/m. The pH values ranged between 7.10 and 7.40 in the successive layers of the studied profiles. This unit comprises 5.68% of El-Galaba basin.
- Soil mapping unit 2 (SMU2): This mapping unit is deep, sandy to sandy loam, non-saline to saline, with ripples, and few rills on the surface. It occupies an area of 1044.00 km^2 (26.56%). The calcium carbonate content ranges between 2.29 and 6.25% in the different layers of the representative profiles. The organic matter content ranges between 0.10 and 0.50%. The soil salinity values reveal that, the electrical conductivity (ECe) ranges between 0.2 and 5.86 dS/m. The pH values ranged between 7.44 and 8.10 in the successive layers of the studied profiles. This unit occupies 26.56% of El-Galaba basin.
- Soil mapping unit 3 (SMU3): This mapping unit is moderately deep, gravely sandy to silty clay, non-saline, with scattered stones and boulders on the surface. It occupies an area of 318.37 km^2 (8.10%). The calcium carbonate content ranges between 3.25 and 7.29% in the different layers of the representative profiles. The organic matter content is very low and ranges between 0.07 and 0.5%. The soil salinity values reveal that, the electrical conductivity (ECe) ranges

between 0.26 and 0.45 dS/m. The pH values ranged between 7.40 and 7.79 in the successive layers of the studied profiles. This unit comprises 8.10% of the study area.

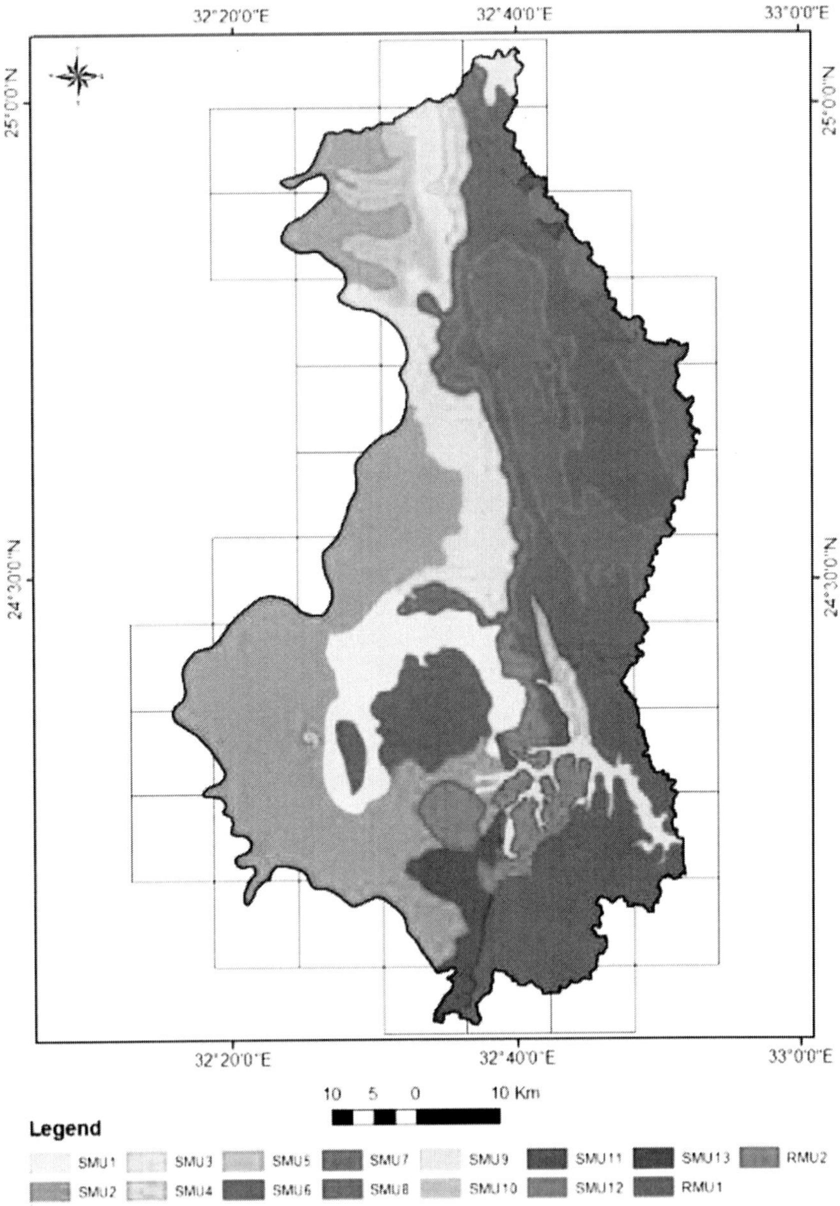

Map 4. Mapping units of El-Galaba basin (Sin El-Kadab plateau was excluded).

- Soil mapping unit 4 (SMU4): This mapping unit is moderately deep, sandy loam, non-saline, with some gravels on the surface. It occupies an area of 65.35 km^2 (1.66%). The calcium carbonate content ranges between 4.42 and 6.25% in the different layers of the representative profiles. The organic matter content is very low and ranges between 0.45 and 0.42%. The soil salinity values reveal that, the electrical conductivity (ECe) ranges between 0.33 and 0.73 dS/m. The pH values ranged between 7.10 and 7.97 in the successive layers of the studied profiles. This unit comprises 1.66% of the study area.
- Soil mapping unit 5 (SMU5): This mapping unit is deep, sandy, non-saline, with gravels and few boulders on the surface. It occupies an area of 180.37 km^2 (3.33%). The calcium carbonate content ranges between 1.88 and 6.67% in the different layers of the representative profiles. The organic matter content is very low and ranges between 0.05 and 0.5%. The soil salinity values reveal that, the electrical conductivity (ECe) ranges between 0.23 and 0.83 dS/m. The pH values ranged between 7.40 and 7.95 in the successive layers of the studied profiles. This unit comprises 3.33% of the study area.
- Soil mapping unit 6 (SMU6): This mapping unit is deep, non-saline to saline, with gravels, stones, and some boulders on the surface. It occupies an area of 526.55 km^2 (13.39%). The calcium carbonate content ranges between 0.50 and 5.21% in the different layers of the representative profiles. The organic matter content is very low and ranges between nil and 0.57%. The soil salinity values reveal that, the electrical conductivity (ECe) ranges between 0.20 and 10.40 dS/m. The pH values ranged between 7.15 and 8.18 in the successive layers of the studied profiles. This unit comprises 13.39% of the study area.
- Soil mapping unit 7 (SMU7): This mapping unit is moderately deep, very saline, with some gravels and boulders on the surface. It occupies an area of 88.95 km^2 (2.26%). The calcium carbonate content ranges between 0.10 and 0.21% in the different layers of the representative profiles. The organic matter content ranges between 0.59 and 1.09%. The soil salinity values reveal that, the electrical conductivity (ECe) ranges between 24.60 and 55.00 dS/m. The pH values ranged between 6.90 and 6.97 in the successive layers of the studied profiles. This unit comprises 2.26% of the study area.
- Soil mapping unit 8 (SMU8): This mapping unit is deep, sandy to loamy sand, non-saline, with some gravels and boulders on the surface. It occupies an area of 519.91 km^2 (13.22%). The calcium carbonate content ranges between 0.50 and 5.63% in the different layers of the representative profiles. The organic matter content is very low and ranges between nil and 0.61%. The soil salinity values reveal that, the electrical conductivity (ECe) ranges between 0.25 and

0.73 dS/m. The pH values ranged between 7.30 and 8.46 in the successive layers of the studied profiles. This unit comprises 13.22% of the study area.

- Soil mapping unit 9 (SMU9): This mapping unit is deep, sandy to loamy sand, non-saline to saline. It occupies an area of 95.30 km^2 (2.42%). The calcium carbonate content ranges between 0.50 and 3.90% in the different layers of the representative profiles. The organic matter content is very low and ranges between 0.02 and 0.54%. The soil salinity values reveal that, the electrical conductivity (ECe) ranges between 0.20 and 5.56 dS/m. The pH values ranged between 7.33 and 7.95 in the successive layers of the studied profiles. This unit comprises 2.42% of the study area.
- Soil mapping unit 10 (SMU10): This mapping unit is moderately deep, sandy to loamy sand, very saline, with gravels on the surface. It occupies an area of 2.02 km^2(0.05%). The calcium carbonate content ranges between 0.10 and 0.21% in the different layers of the representative profiles. The organic matter content ranges between 0.59 and 1.09%. The soil salinity values reveal that, the electrical conductivity (ECe) ranges between 24.60 and 55.00 dS/m. The pH values ranged between 6.90 and 6.97 in the successive layers of the studied profiles. This unit comprises 0.05% of the study area.
- Soil mapping unit 11 (SMU11): This mapping unit is deep, sandy to loamy sand, saline, with some gravels on the surface. It occupies an area of 97.12 km^2 (2.47%). The calcium carbonate content ranges between 1.71 and 3.88% in the different layers of the representative profiles. The organic matter content is very low and ranges between 0.08 and 0.81%. The soil salinity values reveal that, the electrical conductivity (ECe) ranges between 6.10 and 12.40 dS/m. The pH values ranged between 7.23 and 7.96 in the successive layers of the studied profiles. This unit comprises 2.47% of the study area.
- Soil mapping unit 12 (SMU12): This mapping unit is deep, non-saline, with gravels, stones and boulders on the surface. It occupies an area of 3.09 km^2 (0.08%). The calcium carbonate content ranges between 2.92 and 9.17% in the different layers of the representative profiles. The organic matter content is very low and ranges between 0.24 and 0.32%. The soil salinity values reveal that, the electrical conductivity (ECe) ranges between 0.28 and 0.34 dS/m. The pH values ranged between 7.28 and 7.37 in the successive layers of the studied profiles. This unit comprises 0.08% of the study area.
- Soil mapping unit 13 (SMU13): This mapping unit is moderately deep, sandy to loamy sand, non-saline. It occupies an area of 9.51 km^2 (0.24%). The calcium carbonate content ranges between 1.04 and 3.02% in the different layers of the representative profiles. The organic matter content

is very low and ranges between 0.10 and 0.24%. The soil salinity values reveal that, the electrical conductivity (ECe) ranges between 0.19 and 0.27 dS/m. The pH values ranged between 6.45 and 7.56 in the successive layers of the studied profiles. This unit comprises 0.24% of the study area.

- Rocky non-soil mapping unit 1 (RMU1): This mapping unit is severely to weakly dissected plateau rock land.
- Rocky non-soil mapping unit 2 (RMU2): This mapping unit is severely to weakly dissected plateau rock land, denuded, smoothened relief, and with many windblown sand accumulations.

Land use capability

The land use capability modeling resulted in three land use capability classes S2 (Good), S3 (Moderate); and N (Marginal). (Map 5) shows the land use capability resulted from the aforementioned model.

Land with good use capability: It occupies an area of 130.87 km^2 (3.33%). Sandy textured soils with relatively low salinity. Land included in this class has certain topographic and edaphic limitations, which somewhat reduce the set of possible crops and the productive capability. With an appropriate management, this type of land should be of good productivity. Moderate soil conservation practices are required to prevent this land from deterioration.

Land with moderate use capability: It occupies an area of 2990.28 km^2 (76.06%). Deep to moderately deep, gravely sandy to silty clay textured soils and non-saline to very saline land included in this class has considerable limitations linked to topographic, edaphic, or climatic factors. These substantially reduce the range of possible crops and the productive capability. This type of land needs necessarily intensive and special conservation practices to maintain continued productivity. Management techniques are difficult to apply without higher costs.

Marginal or non-productive land: It occupies an area of 3.09 km^2 (0.08%). Deep soils with gravels, stones and boulders on the surface: Land included in this class does not generally provide the ecological conditions necessary for agricultural crops, and is recommended for pasture or forestry as the only way of maintaining and recovering the productive capability. Management and conservation practices of this type of land vary depending on its topographic, edaphic, or climatic deficiencies. Totally non-productive land is included in this class.

Cropping pattern

The physical and chemical soil properties (soil suitability criteria) were further evaluated to define the land suitability for ten land uses which are: wheat, corn, melon, potatoes, sunflower, sugar beet, Alfalfa, peach, citrus, and olive.

For each land use, crop requirements were matched with soil characteristics, and the optimum cropping pattern was obtained for each land utilization type. The land suitability modeling results are shown in (Map 6) for the selected land uses.

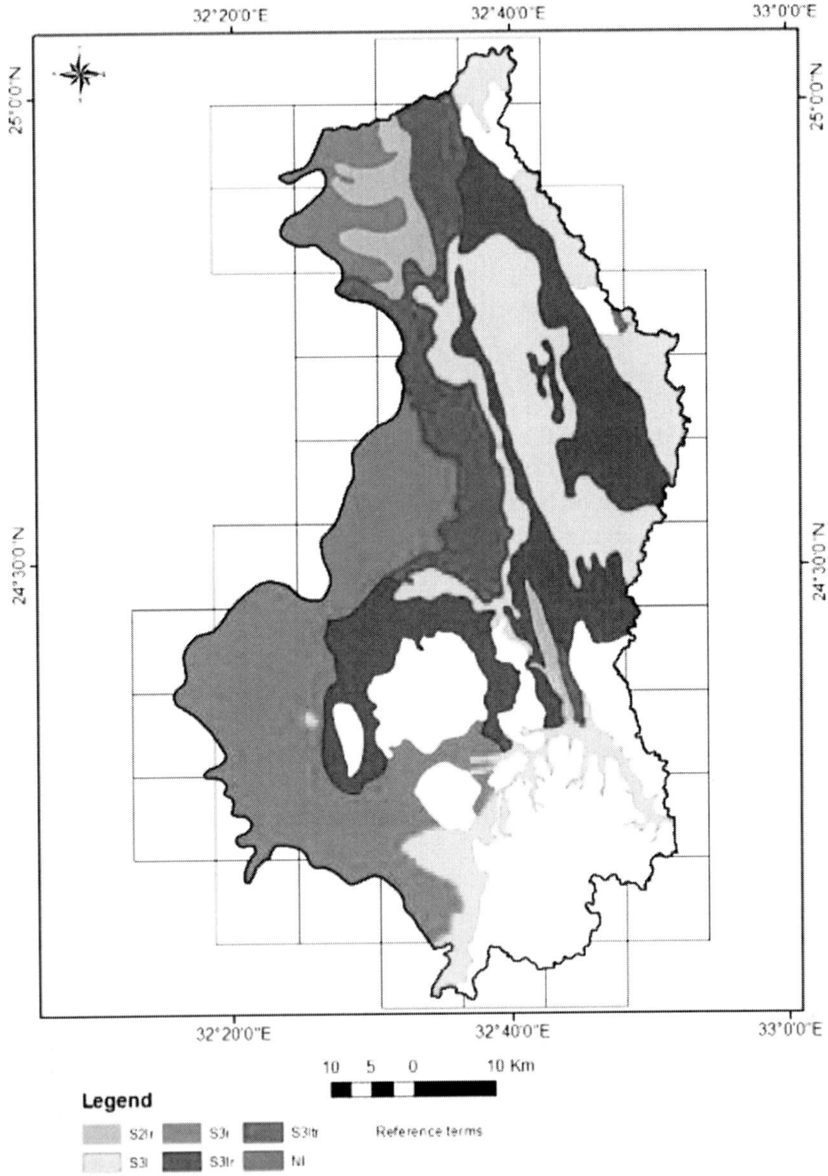

Map 5. Capability classes of El-Galaba basin.

Land with good use capability: It occupies an area of 130.87 km^2 (3.33%). Sandy textured soils with relatively low salinity. Land included in this class has certain topographic and edaphic limitations, which somewhat reduce the set of possible crops and the productive capability. With an appropriate management, this type of land should be of good productivity. Moderate soil conservation practices are required to prevent this land from deterioration.

Land with moderate use capability: It occupies an area of 2990.28 km^2 (76.06%). Deep to moderately deep, gravely sandy to silty clay textured soils and non-saline to very saline land included in this class has considerable limitations linked to topographic, edaphic, or climatic factors. These substantially reduce the range of possible crops and the productive capability. This type of land needs necessarily intensive and special conservation practices to maintain continued productivity. Management techniques are difficult to apply without higher costs.

Marginal or non-productive land: It occupies an area of 3.09 km^2 (0.08%). Deep soils with gravels, stones and boulders on the surface: Land included in this class does not generally provide the ecological conditions necessary for agricultural crops, and is recommended for pasture or forestry as the only way of maintaining and recovering the productive capability. Management and conservation practices of this type of land vary depending on its topographic, edaphic, or climatic deficiencies. Totally non-productive land is included in this class.

Cropping pattern

The physical and chemical soil properties (soil suitability criteria) were further evaluated to define the land suitability for ten land uses which are: wheat, corn, melon, potatoes, sunflower, sugar beet, Alfalfa, peach, citrus, and olive. For each land use, crop requirements were matched with soil characteristics, and the optimum cropping pattern was obtained for each land utilization type. The land suitability modeling results are shown in (Map 6) for the selected land uses.

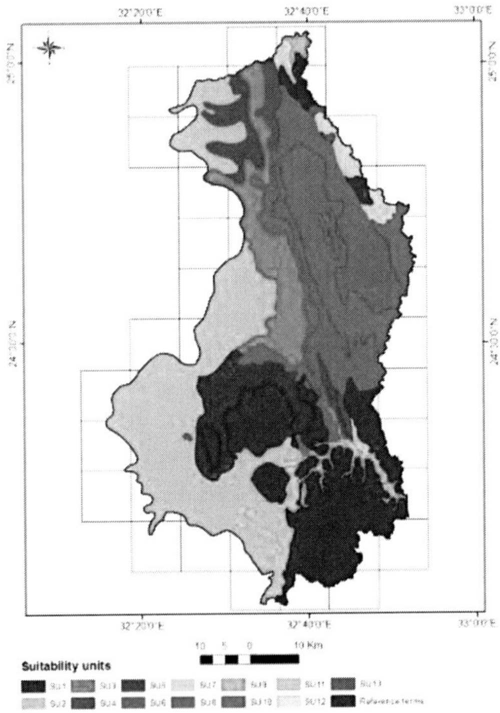

Suitability unit	Land use									
	Wheat	Corn	Melon	Potatoes	Sunflower	Sugar beet	Alfalfa	Peach	Citrus	Olive
SU1	S2c	S1	S2t	S2t	S2c	S2ca	S2c	S2pt	S2pt	S3t
SU2	S4t	S4t	S3t	S3t	S4t	S4t	S4t	S3t	S3t	S3t
SU3	S5t	S5t	S5t	S5t	S5t	S5t	S5t	S5t	S5t	S4t
SU4	S5t	S5t	S5t	S5t	S5t	S5t	S5t	S4t	S4t	S4t
SU5	S3t	S3t	S3t	S3t	S3t	S3t	S3t	S2td	S2td	S2tdc
SU6	S3t	S3t	S3t	S3t	S3t	S3t	S3t	S5t	S5t	S5t
SU7	S4t	S4t	S4t	S4t	S4t	S4t	S4t	S3t	S3t	S3t
SU8	S3t	S3t	S2ts	S2ts	S3t	S3t	S3t	S2ts	S2ts	S2ts
SU9	S3t	S3t	S2t	S2t	S3t	S3t	S3t	S2t	S2t	S2tc
SU10	S4t	S4t	S4t	S4t	S4t	S4t	S4t	S3t	S3t	S3t
SU11	S3t	S3t	S2t	S2t	S3t	S3t	S3t	S2t	S2t	S2tc
SU12	S5s	S5s	S5s	S5s	S5s	S4s	S4s	S5s	S5s	S5s
SU13	S4t	S4t	S4t	S4t	S4t	S4t	S4t	S4s	S4s	S3ts

* Suitability classes: (S1) soils with optimum suitability – (S2) soils with high suitability – (S3) soils with moderate suitability – (S4) soils with marginal suitability – (S5) soils with no suitability.
* Soil limitations: (p) useful depth - (t) texture - (d) drainage condition - (c) carbonates content - (s) –salinity - (a) sodium saturation.

Map 6. Combined suitability unit and suitability classes for the selected land uses in El-Galaba basin.

CONCLUSION

The objective of this study was to investigate the potentiality of El-Galaba basin soils, which is a part of the Western Desert of Egypt and located in the western part of the Aswan Governorate, for the horizontal agricultural expansion and their optimum agricultural use. Remote sensing and GIS are very helpful tools to store, manipulate and quantitatively evaluate soil capability and suitability. The results of the study revealed that about 79.39% of El-Galaba basin area is good to moderately good of land use capability. The main capability limitations were soil and erosion risks. Also, the suitability analyses show that the basin is suitable for ten land uses which are: wheat, corn, melon, potatoes, sunflower, sugar beet, Alfalfa, peach, citrus, and olive. El-Galaba basin is of high potentiality for horizontal agricultural expansion. This area will benefit if development is planned and executed in a manner that takes advantage of the natural resources without threatening their quality.

ACKNOWLEDGMENTS

The authors acknowledge the National Authority for Remote Sensing and Space Sciences (NARSS) for funding and providing all the supporting material for an internal research project principally investigated by the first author of the present work, from which data of the present work were derived. ASTER GDEM is a product of The Ministry of Economy, Trade and Industry of Japan (METI) and the National Aeronautics and Space Administration (NASA).

REFERENCES

1. Abd El-Razik, T., Razavaliaev, A., 1972. On the tectonic origin of the Nile Valley between Idfu and Qena, Egypt. J. Geol. 16 (2), 235–245.
2. Attia, M.I., 1954. Deposits in the Nile Valley and the Delta, Geol. Survey of Egypt. Government Press, Cairo. Brookes, I.A., 2003.
3. Geomorphic indicators of Holocene winds in Egypt's Western Desert. Geomorphology 56, 155–166.
4. Clements, R., 2009. The economic cost of climate change in Africa. http://www.christianaid.org.uk/images/economic-cost-of-climatechange-in-frica.pdf.

5. De la Rosa, D., Magaldi, D., 1992. Methodologies in evaluation system for de tierras para Mediterranean region. Soc. Esp. Cien. Suelo. Madrid.
6. De la Rosa, D., Mayol, F., Diaz-Pereira, E., Fernandez, M., De la Rosa, D.J., 2004. A land evaluation decision support system (MicroLEIS DSS) for agricultural soil protection with special reference to the Mediterranean region. Environ. Modell. Software 19, 929–942.
7. De la Rosa, D., Moreno, J.A., Garcia, L.V., Almorza, J., 1992. MicroLEIS: a microcomputer-based Mediterranean land evaluation information system. Soil Use Manag. 8, 89–96.
8. Dent, D., Young, A., 1981. Soil survey and land evaluation. George Allen & Unwin, London. ESRI, 2001. Arc-GIS Spatial Analysis: Advanced GIS Spatial Analysis Using Raster and Vector Data. ESRI, 380, New York, CA92373-8100 USA.
9. ESRI, 2013. ArcGIS Desktop: Release 10.2. Redlands, CA, USA. Exelis Visual Information Solutions, 2012. ENVI Software Version 5.0, Boulder, Colorado.
10. Food and Agriculture Organization of United Nations, 1966. High Dam Soil Survey Project. Food and Agriculture Organization of United Nations, Ministry of Agriculture, Egypt. FAO, 1976. A framework for land evaluation. Soils Bulletin 32. Food and Agriculture Organization of the United Nations, Rome.
11. FAO, 1985. Land evaluation for irrigated agriculture. Soils Bulletin No. 55, FAO, Rome.
12. FAO, 1991. Land use planning applications. Bulletin No. 68, FAO, Rome.
13. FAO, 2003. Theoretical framework for land evaluation. Geoderma 72, 165–190, FAO, Rome.
14. FAO, 2006. Guidelines for Soil Profile Description, third ed. Soil resources management and conservation service, land and water development division, FAO, Rome
15. FAO, 2007. Land evaluation, towards a revised framework. FAO, Rome.
16. Foote, K.E., Lynch, M., 1996. Geographic information systems as an integrating technology: context, concepts and definition. University of Texas, Austin.
17. Ghafari, A., Cook, H.F., Lee, H.C., 2000. Integrating climate, soil and crop information: a land suitability study using GIS. In: 4th International Conference on Integrating GIS and Environmental Modeling (GIS/EM4). Problems, Prospects and Research Needs, Banf, Alberta.
18. Goudie, A.S., 2002. Great warm desert of the world: landscapes and evolution. Oxford University Press, New York, p 444. Issawi, B., 1983. Ancient rivers of the Eastern Egyptian Desert. Episodes 2, 3–6.

19. Issawi, B., Hinnawi, M., 1980. Contribution to the geology of the Plain West of the Nile between Aswan and Kom Ombo. In: Wendorf, F., Schild, R. (Eds.), Loaves and Fishes: The Prehistory of Wadi Kubbaniya. Southern Methodist University Press, Dallas, pp. 311–330.
20. Issawi, B., McCauly, J.F., 1993. The Cenozoic landscape of Egypt and its river systems. Ann. Geol. Surv. Egypt XIX, 357–384.
21. Jenson, S.K., Domingue, J.O., 1988. Extracting topographic structure from digital elevation data for geographic information system analysis. Photogramm. Eng. Rem. S. 54 (11), 1593–1600.
22. Koch, M., Gaber, A., Burkholder, B., Geriesh, M., 2012. Development of new water resources in Egypt with earth observation data: opportunities and challenges. Int. J. Environ. Sustainability 1 (3), 1–11.
23. Lillesand, T.M., Kiefer, R.W., 2000. Remote sensing and image interpretation. John Wiley & Sons, New York. Mohamed, E.S., Belal, A.A., Saleh, A.M., 2013. Assessment of land degradation east of the Nile Delta, Egypt, using remote sensing and GIS techniques. Arabian J. Geosci. 6 (8), 2843–2853.
24. Mohamed, E.S., Saleh, A.M., Belal, A.A., 2014. Sustainability indicators for agricultural land use based on GIS spatial modeling in North Sinai – Egypt. Egypt. J. Remote Sens. Space Sci. 17, 1–15.
25. Nelson, G.C., Rosegrant, M.W., Koo, J., Robertson, R., Sulser, T., Zhu, T., Ringler, C., Msangi, S., Palazzo, A., Batka, M., Magalhaes, M., Valmonte-Santos, R., Ewing, M., Lee, D., 2009.
26. Climate change impact on agriculture and costs of adaptation. International Food Policy Research Institute, Washington, DC. ONERN, 1982. Clasificacio´n de las tierras Del Peru´. Oficina Nacional de Evaluacio´n de, Lima – Peru´. Palacios-Orueta, A.P., Ustin, S.L., 1998. Remote sensing of soil properties in the Santa Monica mountains, spectral analysis. Remote Sens. Environ. 65, 170–183.
27. Said, R., 1962. The Geology of Egypt. Elsevier Publishing Company, Amsterdam – New York.
28. Saleh, A.M., Belal, A.A., Arafat, S.M., 2013. Identification and mapping of some soil types using field spectrometry and spectral mixture analyses: a case study of North Sinai, Egypt. Arabian J. Geosci. 6 (6), 1799–1806.
29. Saleh, A.M., Belal, A.A. , 2014. Delineation of site-specific management zones by fuzzy clustering of soil and topographic attributes: a case study of East Nile Delta, Egypt. In: 8th International Symposium of the Digital Earth (ISDE8). IOP Conf. Series: Earth and Environmental Science 18 (2014) 012046. IOP Publishing.
30. Salman, A.B., 1974. Structural and radioactivity of some phosphate deposits, East Luxor area, Egypt (Ph.D. thesis). Ain Shams University, Cairo, Egypt.

31. Soil Survey Staff, 1999. Soil taxonomy: a basic system of soil classification for making and interpreting soil surveys, second ed. Natural Resources Conservation Service. U.S. Department of Agriculture Handbook 436.
32. Soil Survey Staff, 2014. Soil Survey Field and Laboratory Methods Manual. Soil Survey Investigations Report No. 51, Version 2.0. Manual, Natural Resources Conservation Service, U.S. Department of Agriculture.
33. Strahler, A.N., 1957. Quantitative analysis of watershed geomorphology. Trans. Am. Geophys. Union 8 (6), 913–920.
34. Sys, C.E., Van Ranst, E., Debaveye, J., Beernaert, F., 1993. Land evaluation. Part III: crop requirements. Agricultural Publications n 7. General Administration for Dev. Co-operation, Brussels, Belgium.
35. Tarboton, D.G., Bras, R.L., Rodriguez-Iturbe, I., 1991. On the extraction of channel networks from digital elevation data. Hydrol. Process. 5, 81–100.
36. USGS, 2013. Using the USGS Landsat 8 Product. Retrieved from USGS: http://landsat.usgs.gov/Landsat8_Using_Product.php.
37. Van Diepen, C.A., Van Keuken, H., Wolf, J., Berkhout, J.A.A., 1991. Land evaluation from intuition to quantification. Advances in Soil Science, vol. 15. Springerm, New York.
38. Verheye, W., 1986. Land evaluation and land use planning in The EC. Proc. EC Land Eval. Work, Brussels. Wood, J., 1996. The Geomorphological Characterization of Digital Elevation Models (Ph.D. thesis). University of Leicester, Department of Geography, Leicester, UK.
39. World Food Programme, 2013. The Status of Poverty and Food Security in Egypt: Analysis and Policy Recommendations. World Food Programme.
40. Youssef, M.I., 1968. Structural pattern of Egypt and its interpretation, AAPG Bulletin.

CITATION

A.M. Saleh, A.B. Belal, E.S. Mohamed, Land resources assessment of El-Galaba basin, South Egypt for the potentiality of agriculture expansion using remote sensing and GIS techniques, The Egyptian Journal of Remote Sensing and Space Science, Available online 31 July 2015, ISSN 1110-9823, http://dx.doi.org/10.1016/j.ejrs.2015.06.006.

CHAPTER 4

Machine Learning in Geosciences and Remote Sensing

David J. Lary[1], Amir H. Alavi[2], Amir H. Gandomi[3], Annette L. Walker[4]

[1] Hanson Center for Space Science, University of Texas at Dallas, Richardson, TX 75080, USA
[2] Department of Civil and Environmental Engineering, Michigan State University, East Lansing, MI 48824, USA
[3] BEACON Center for the Study of Evolution in Action, Michigan State University, East Lansing, MI 48824, USA
[4] Aerosol and Radiation Section, Naval Research Laboratory, 7 Grace Hopper Ave., Stop 2, Monterey, CA 93943-5502, USA

Highlights

- Highlighting the role of machine learning for solving problems in geosciences and remote sensing.
- Developing a typical nonparametric regression for the abundance of ground level airborne particulate matter.
- Developing a typical classification model for identifying global dust sources.

ABSTRACT

Learning incorporates a broad range of complex procedures. Machine learning (ML) is a subdivision of artificial intelligence based on the biological learning process. The ML approach deals with the design of algorithms to learn from machine readable data. ML covers main domains such as data mining, difficult-to-program applications, and software applications. It is a collection of a variety of algorithms (e.g. neural networks, support vector machines, self-organizing map, decision trees, random forests, case-based reasoning, genetic programming, etc.) that can provide multivariate, nonlinear, nonparametric regression or classification. The modeling capabilities of the ML-based methods have resulted in their extensive applications in science and engineering. Herein, the role of ML as an effective approach for solving problems in geosciences and remote sensing will be highlighted. The unique features of some of the ML techniques will be outlined with a specific attention to genetic programming paradigm. Furthermore, nonparametric regression and classification illustrative examples are presented to demonstrate the efficiency of ML for tackling the geosciences and remote sensing problems.

INTRODUCTION

Machine learning (ML) is an effective empirical approach for both regression and/or classification (supervised or unsupervised) of nonlinear systems. Such systems can be massively multivariate involving a few or literally thousands of variables. In ML, a comprehensive 'training dataset' of examples is constructed covering as much of the system parameter space as possible. Typically, a random subset of the data is put aside for a completely independent validation. ML is ideal for addressing those problems where our theoretical knowledge is still incomplete but for which we do have a significant number of observations and other data. In an ideal world, if we had complete theoretical understanding, ML would be superfluous.

ML has proven useful for a very large number of applications in many parts of the earth system (land, ocean, and atmosphere) and beyond, from retrieval algorithms, crop disease detection, new product creation, bias correction and code acceleration (e.g. Yi and Prybutok, 1996, Atkinson and Tatnall, 1997, Carpenter et al., 1997, Lary et al., 2004 and Lary et al., 2009; Brown et al. 2008; Azamathulla, 2012, Zahabiyoun et al., 2013 and Madadi et al., 2015). The types of the ML algorithms commonly used are artificial neural networks (ANN), support vector machines (SVM), self-organizing map (SOM), decision trees (DT), ensemble methods such as random forests, case-based reasoning, neuro-fuzzy (NF), genetic algorithm (GA), multivariate adaptive regression splines (MARS), etc (e.g., Shahin et al., 2001, Shahin and Jaksa, 2005, Das and Basudhar, 2008, Samui, 2008a, Samui, 2008b, Samui, 2012, Azamathulla and Wu, 2011, Azamathulla et al., 2011, Azamathulla et al., 2012, Garg et al., 2014a, Garg et al., 2014b and Garg et al., 2014c). The ML-based methods have been widely applied to the science and engineering problems for near two decades. This is while the application of these techniques in the geosciences and remote sensing area is fairly new and limited. Herein, a number of relevant and documented applications of ML will be summarized. The unique features of some of the ML techniques for dealing with the geosciences and remote sensing problems will be reviewed. Moreover, two very different but complementary illustrative examples are presented: one using multivariate nonlinear nonparametric regression, and the other using multivariate nonlinear unsupervised classification. For these two illustrative cases, we will start with the scientific motivation that makes clear the real need for ML and then demonstrate how ML addresses this need.

OVERVIEW OF ML APPLICATIONS IN GEOSCIENCES AND REMOTE SENSING

The ML algorithms are "universal approximators". That is, they learn the underlying behavior of a system from a set of training data. Another interesting feature of the ML-based techniques is that they do not need a prior knowledge about the nature of the relationships between the data. The application of ML may be categorized into three areas (Lary, 2010):

(1) The system's deterministic model is computationally expensive and ML can be used as a code accelerator tool.
(2) There is no deterministic model but an empirical ML-based model can be derived using the existing data.
(3) Classification problems.

As mentioned before, ML includes a variety of algorithms ANN, SVM, SOM, and DT. Over the last decade, there has been considerable progress in developing ML-based methodologies for many of Earth Science applications (Lary, 2010). Some of these studies have received special recognition as a NASA Aura Science highlight (Lary et al., 2007) and commendation from the NASA MODIS instrument team (Lary et al., 2009). ANN and SVM are the most commonly used ML techniques for dealing with geoscience problems. A comprehensive review of application of ANN and SVM in geoscience and remote sensing can be found in Lary (2010). Also, Nikravesh (2007) presented an inclusive review study of the application of neuro-computing, fuzzy logic and evolutionary computing in geosciences and oil exploration. That study also covers the successful application of hybrid methodologies such as NF, neural-genetic, fuzzy-genetic and neural-fuzzy-genetic in the field. Nikravesh (2007) discussed the major impact of these techniques for tackling problems in geophysical, geological and reservoir engineering (e.g., intelligent reservoir characterization and exploration, seismic data processing, and characterization, well logging, reservoir mapping, etc.).

Among the main subsets of ML, applications of genetic programming (GP) (Koza, 1992) in the geoscience and remote sensing domain are very new and restricted to a few areas. Despite the good performance of ANNs, SVM and many of the other ML methods, they are considered as black-box models. That is, they are not capable of generating practical prediction equations. GP is considered as an efficient approach to deal with this issue. GP uses the principle of Darwinian natural selection to generate computer programs for solving a problem. In fact, GP is a specialization of GA where the encoded solutions (individuals) are computer programs rather than binary strings (Alavi and Gandomi, 2011). A notable feature of GP and its variants is that they can produce prediction equations without a need to pre-define the form of the existing relationship (Alavi et al., 2010, Alavi and Gandomi, 2011, Alavi et al., 2011a and Gandomi and Alavi, 2011). Herein, we present an overview of a number of relevant and recent applications of GP in the field. The majority of applications of GP focus on the behavioral characterization of rock mass. The other few studies use GP as a tool for interpreting the remote sensing data. It is worth mentioning that there are some other studies mainly on the applicability of GP for analyzing geotechnical engineering problems such as liquefaction phenomenon, ground motion parameters, or ground movement patterns (e.g., Javadi et al., 2006, Shuhua et al., 2006, Lia et al., 2007, Cabalar and Cevik, 2009, Gandomi et al., 2011, Alavi et al., 2011b, Alavi and Gandomi, 2012 and Gandomi and Alavi, 2013).

As mentioned before, most of the GP-based studies focus on estimating the properties of rock. Perhaps, one of the pioneer studies in the field was done by Baykasoglu et al. (2008). They applied GP-based approaches to the strength prediction of limestone. Different variants of GP, called multi expression programming (MEP), gene expression programming (GEP) and linear genetic programming (LGP) to the uniaxial compressive strength (UCS) and tensile strength prediction of chalky and clayey soft limestone. The models were developed using experimental data. The models had a good accuracy with determination coefficient (R^2) equal to 0.76 and 0.95 for tensile strength and UCS, respectively. Beiki et al. (2010) developed new models to determine the deformation modulus of rock masses using GP. Several parameters were used as the predictor variables such as modulus of elasticity of intact rock (Ei), uniaxial compressive strength (UCS), rock mass quality designation (RQD), the number of joint per meter (J/m), porosity, dry density, and geological strength index

(GSI). Beiki et al. (2010) also found that the GP models give higher predictions over existing empirical models. Recently,Karakus (2011) employed GP to analyze laboratory strength and elasticity modulus data for some granitic rocks. Uniaxial compressive strength (σ_c), tensile strength (σ_t) and elasticity modulus (E) were formulated in terms of total porosity (n), sonic velocity (V_p), point load index (Is) and Schmidt Hammer values (SH). The results clearly indicated that GP is a potential tool for predicting the elasticity modulus and the strength of granitic rocks.

Rock mass modulus of deformation (E_m) plays a critical role in designing many structures on rock. Ravandi et al. (2013) performed a back analysis calculation to derive an equation for estimation of E_m using GP. The model was developed using a database of 40,960 datasets, including vertical stress (r_z), horizontal to vertical stresses ratio (k), Poisson's ratio (m), radius of circular tunnel (r) and wall displacement of circular tunnel on the horizontal diameter (d). The computer program (CP) generated by GP had a good accuracy with a correlation coefficient equal to 0.97. More recently, Ozbek et al. (2013)proposed models to estimate the UCS of rocks with different characteristics using a GP branch, i.e., GEP. They have considered five different types of rocks including basalt and ignimbrite (black, yellow, gray, brown) were prepared. UCS was formulated in terms of effective porosity (n), water absorption by weight (w_A), and unit weight (γ). It was shown that GP can be used for estimating the UCS of rocks successfully.

The ML-based techniques are increasingly used for interpreting the remote sensing images (RSIs). Conversely from the other ML methods, there are few GP-based studies in the field of remote sensing technology. Some typical examples are estimation of the typhoon rainfall over ocean using multi-variable meteorological satellite data (Chen et al., 2011), monitoring reservoir water quality using remote sensing images (Chen, 2003), mapping of base-metal deposits (Lewkowski et al., 2010), image thresholding for landslide detection (Rosin and Hervas, 2002), and soil moisture distribution analysis (Makkeasorn et al., 2006). As good examples in this context, let us consider the studies done by Makkeasorn et al. (2009) and dos Santos et al. (2010). RSIs are widely used as valuable tools in different real world applications. In the context of agribusiness applications, a major challenge is recognition of crop type regions. To cope with this issue, dos Santos et al. (2010) proposed a new GP-based approach for automatic recognition of coffee crops in RSIs. They combined texture and spectral information encoded by image descriptors. Fig. 1 shows the steps of the proposed classification process. As it is seen, this approach can be divided into two main phases: (1) the image description and (2) image classification. The first phase including the Step 1 to 3 is focused on the image content characterization. The remaining 4 steps belong to the image classification process. GP has been used by dos Santos et al. (2010) to identify relevant partitions by combining the similarities provided by descriptors. Later, dos Santos et al. (2010) proved that their GP-based method yields slightly better results than the traditional MaxVer approach.

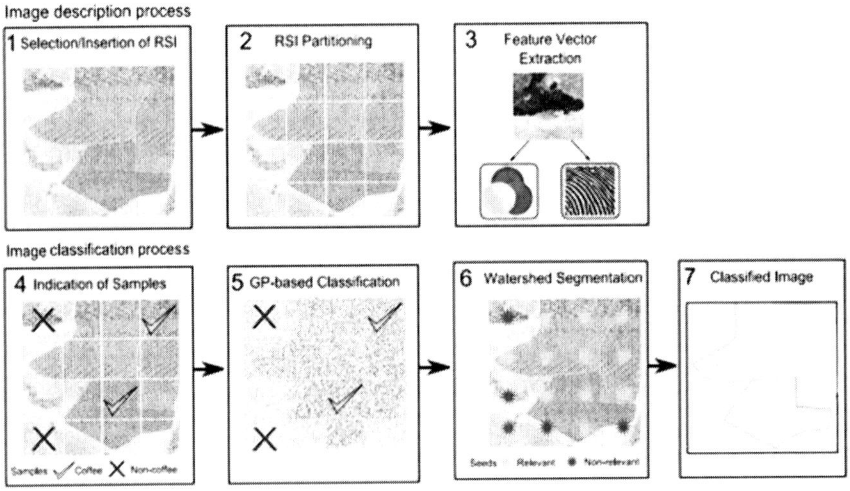

Figure 1. Steps of the classification process (dos Santos et al., 2010).

The other example is about the detection of seasonal change of riparian zones with remote sensing images and GP. As it is known, riparian zones have a notable impact on the maintenance of ecosystem integrity region wide. In order to detect the seasonal change of riparian zones, Makkeasorn et al. (2009) developed a GP-based method, called the RIparian Classification Algorithm (RICAL). This approach incorporates vegetation indices and soil moisture images derived from LANDSAT 5 TM and RADARSAT-1 satellite images, respectively. Makkeasorn et al. (2009) estimated the soil moisture based on RADARSAT-1 Synthetic Aperture Radar (SAR) images and by using GP. Then, they defined several vegetation indices based on reflectance factors calculated as the response of the instrument on LANDSAT. The defined spectral indices along with soil moisture images were used to classify the riparian zones through another GP analysis.

ILLUSTRATIVE EXAMPLES

In order to demonstrate the efficiency of ML for tackling the geosciences and remote sensing problems, two nonparametric regression and classification illustrative examples are presented in this section.

Characterizing airborne particulates

Climate change is a defining issue of our generation. The IPCC have reported that one of the largest uncertainties in simulating climate change is the radiative forcing associated with atmospheric aerosols (IPCC, 2013). In addition, the World Health Organization (WHO, 2014) just released a report which concluded that in 2012, seven million deaths across the world were associated with air pollution, with a significant contribution from atmospheric aerosols. However, due to the significant challenges of remotely sensing the atmospheric boundary layer, we still do not have a definitive aerosol abundance product for the boundary layer. We demonstrate that by simultaneously combining around 50 massive remote sensing and Earth System modeling products using multivariate nonparametric nonlinear ML, we can create a Virtual Sensor providing an accurate boundary layer atmospheric aerosol product together with an associated uncertainty. We can validate the product using a global sensor web of ground based sensors and unmanned aerial vehicles. The approach is also of benefit for producing societally relevant data products.

Atmospheric aerosols are found globally. Estimating the abundance of airborne particulate matter is a critical yet challenging task for both retrospective study and forecasting of air quality (Grell et al., 2005 and Chuang et al., 2012), visibility and climate change (Hansen et al., 1988 and Allen et al., 2000).

Numerous studies show that among air pollutants, the abundance of ground level airborne particulate matter with a diameter of 2.5 μm or less ($PM_{2.5}$) has the strongest link with human health effects (Brook et al., 2010). Increased morbidity and mortality has been associated with exposure to $PM_{2.5}$ thereby suggesting improved life expectancy is possible by reducing the exposure level (Pope et al., 2009). Not only in the US but also in European studies, a significant number of premature deaths, including cardiopulmonary and lung-cancer deaths were attributed to long-term exposure to $PM_{2.5}$ (Boldo et al., 2011).

For more than half a century researchers have been studying the impact of PM on health. Initially the attempt was to learn about the possible adverse effects, and then the focus shifted to investigate the exposure-response relationships. Now with further advancement in technology and more awareness of health-concerns, studies on composition-specific effects have emerged (Ayala et al., 2012). With implementation of computational fluid dynamics (CFD) models and digital imaging of organs, researchers started to study the pathophysiology associated with PM to better understand the translocation of particulates in human body after their deposition and the fate of these particulates in impacting health.

Most short-term exposure impact studies on $PM_{2.5}$, whether for morbidity or mortality, focus on cardiovascular/cardiopulmonary (Brook et al., 2010) or respiratory (Dockery et al., 1993) conditions. Our dataset, with daily temporal scale, is suitable for such studies. We are already studying daily asthma-related hospital admissions associated with $PM_{2.5}$ using our estimated data.

On the other hand, diseases, such as lung cancer, require study of the long-term exposure to $PM_{2.5}$. Data generated from this study is expected to contribute to Health Impact Assessment (HIA) in different parts of the world concerning long-term exposure to $PM_{2.5}$. Currently, long-term PM values are not available in many localities and in many instances $PM_{2.5}$ values are estimated from PM for long-term HIA (Boldo et al., 2011). Studies also suggest that even low level $PM_{2.5}$ exposure can contribute to serious health impacts (Pope and Dockery, 2006). We have already created daily global estimates of $PM_{2.5}$ with an associated uncertainty for more than 13 years providing an appropriate dataset for extended cohort studies for the areas with both high and level concentrations of ambient $PM_{2.5}$. In addition, long-range transportation of particles as dust can provide potential vectors for bacteria (Ginoux and Torres, 2003 and Prospero, 2003). With global coverage of this study, tracking $PM_{2.5}$ transport is now easier for public health surveillance. Since many of the health conditions are interlinked, comprehensive studies are required to better understand the impact of $PM_{2.5}$. With increasing availability of electronic health records, reliable $PM_{2.5}$ data with seamless temporal and geographic coverage can contribute to revealing many unknowns of $PM_{2.5}$ impacts on health.

It could be noted that the type and degree of adverse effect greatly depends on the composition of the particulate matters. Composition mostly varies due to source materials. Our current study does not provide information on the composition of $PM_{2.5}$. However, this study can be extended to examine the potential of source apportionment considering land use/land cover conditions and transportation mechanism. Recent studies show specific adverse impacts of exposure to ultrafine particles. Future studies are recommended to derive further size fractions beyond just $PM_{2.5}$, particularly the ultra-fine particles in the sub-micron size range.

The increasing awareness of the many health impacts implies the importance of having precise estimations of $PM_{2.5}$. Existing remote sensing data can be effectively used to meet this need by employing ML, ensemble of random forests, to estimate the daily global $PM_{2.5}$ abundance. The method utilizes remote sensing and meteorological data and ground-based observations of particulate matter at 3019 sites in 38 countries (seeFig. 2) (Lary et al., 2014a). Referring to Fig. 2, it can be observed that the sites in North America, Europe and Asia have higher density.

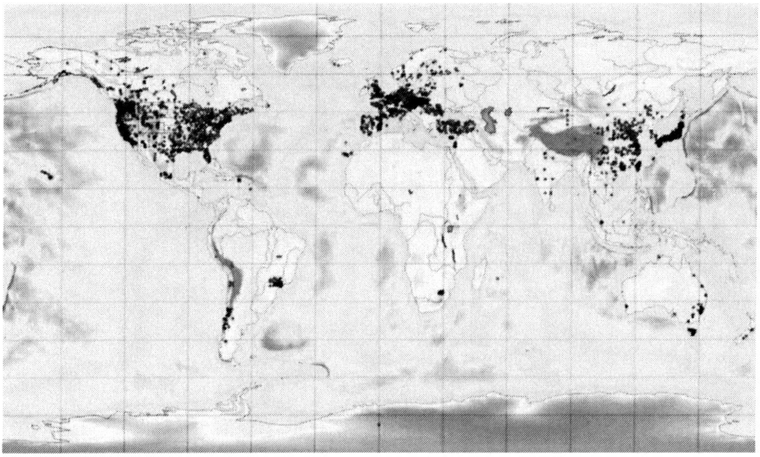

Figure 2. The 8329 $PM_{2.5}$ measurement site locations from 55 countries (red squares) that were used over the period 1997–2014.

The health impacts of $PM_{2.5}$ depend on its abundance at ground level where people can inhale the $PM_{2.5}$. Various networks of ground-based sensors routinely measure the abundance of $PM_{2.5}$. Fig. 2 indicates many gaps in the spatial coverage with no $PM_{2.5}$observations. This is mainly because of the poor coverage of the sensor network. This limitation has been tackled using remote sensing and satellite-derived Aerosol Optical Depth (AOD) coupled with numerical models (Engel-Cox et al., 2004a, Engel-Cox et al., 2004b, Liu et al., 2007, Lee et al., 2011a and Lee et al., 2011b). It has been proved that several parameters such as humidity, temperature, topography, cloud cover, and cloud optical depth affect the relationship between $PM_{2.5}$ and AOD (e.g. Rajeev et al., 2008,Schaap et al., 2009, Liu and Harrison, 2011 and Liu et al., 2012). Thus, a multivariate, non-linear and non-parametric ML approach seems to the best option to capture this relationship. ML has a very good performance in providing a new $PM_{2.5}$ product. Fig. 3presents the monthly average of the ML $PM_{2.5}$ product ($\mu g/m^3$). A good agreement exists between the $PM_{2.5}$ product and the observations. In other words, the color fill of the circles depicting the observations is in good agreement with the background color depicting the new ML $PM_{2.5}$ product.

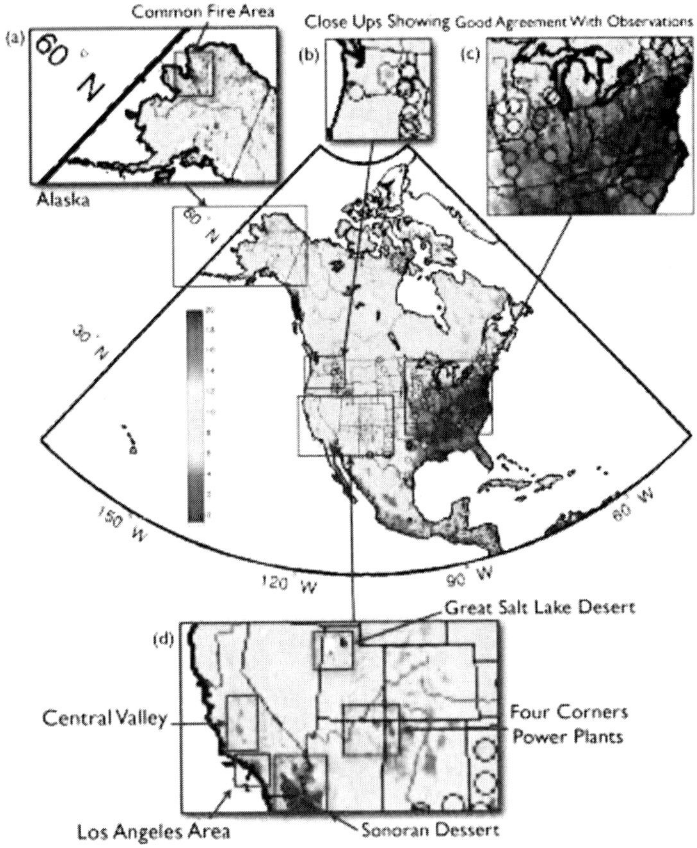

Figure 3. The monthly average ML $PM_{2.5}$ product ($\mu g/m^{-3}$) for August 2001 (Lary et al., 2014a).

Careful attention to details is critical for the precision of $PM_{2.5}$ simulations. First, a *highly restrictive coincident requirement* needs to be considered for the training dataset. Herein, we *only* take into account hourly $PM_{2.5}$ and satellite observations that were made within 30 min of each other and had a great circle distance separation up to only 0.02°. Second, a comprehensive training dataset was gathered spanning the globe for more than a decade. Third, the use of the full range of training parameters that characterize the local environment as carefully as possible is recommended (e.g. humidity, temperature, boundary layer height, surface pressure, etc.). Fourth, a multivariate, nonlinear, nonparametric ML approach is utilized that can handle continuous real variables and categorical variables (flags and masks). It is notable that remotely sensed AOD products typically have inherent biases relative to the ground truth from AERONET and are dependent on the remote sensing instrument, the algorithm, and the version of the processing software (collection). *These biases were directly and individually calibrated for* by performing separate ML training for our target variable ($PM_{2.5}$) for each instrument, software version and algorithm combination (i.e. collection, deep blue, standard). Example of these separate calibrations can be seen in Fig. 4.

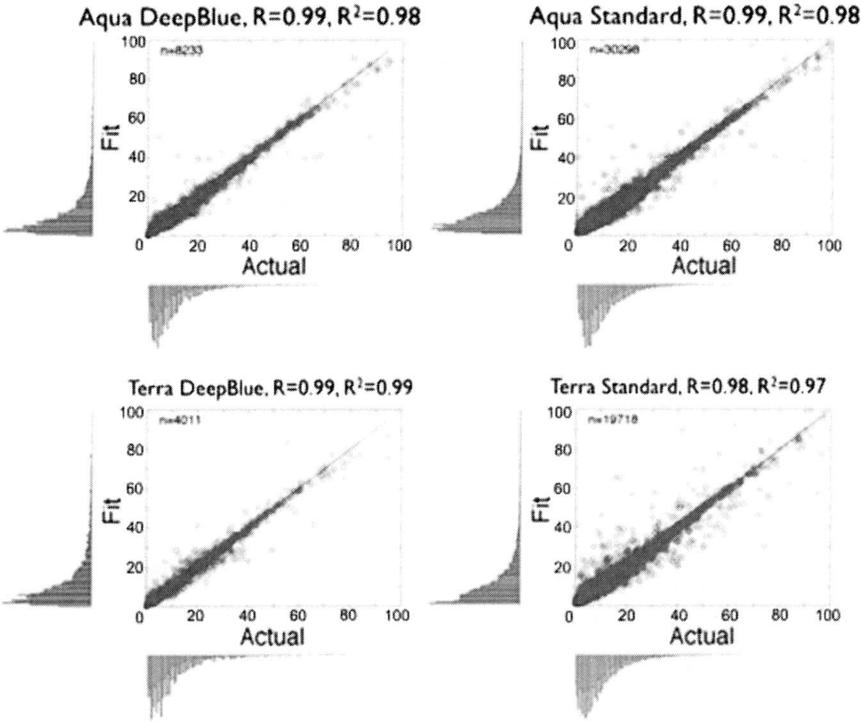

Training dataset statistics and global 2000-2012 correlation coefficients.

	n	R	R^2
Aqua Deep Blue	8,233	0.99	0.98
Aqua Standard	30,298	0.99	0.98
Terra Deep Blue	4,011	0.99	0.98
Terra Standard	19,718	0.98	0.97

Figure 4. The quality of the ML fits is always quantified by scatter diagrams of the observed "truth" plotted on the *x*-axis against the corresponding ML estimate plotted on the *y*-axis. A separate ML fits of $PM_{2.5}$ is performed for each satellite data product using a given algorithm and instrument.

Dust sources

Dust sources of many kinds are found globally. One of the most salient features of dust sources is that they are often very localized. For example, in Figure 5 and Figure 6, we can clearly see that the source of the dust plumes are best described as an ensemble of many *point sources*, *not* broad dust emitting regions. Realistically capturing this very*localized* nature of dust sources has so far largely eluded automated diagnosis, and consequently, description in global models. Invariably current models describe dust sources as rather large scale features (Ginoux et al., 2011), even when vegetation indices and similar approaches are used. This is in marked contrast to what we consistently see in the satellite imagery across the planet.

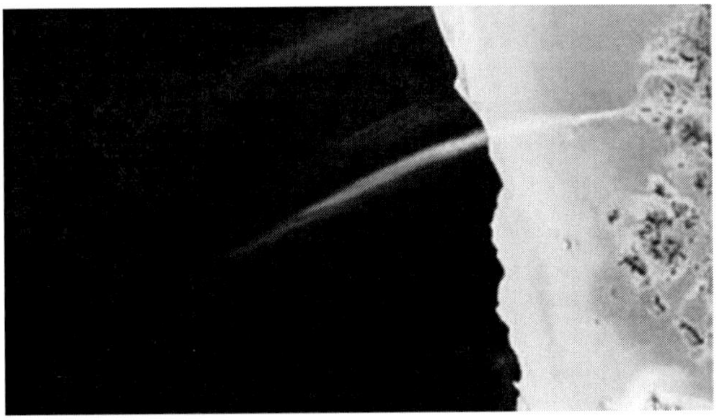

Figure 5. Dust sources are typically localized point sources.

Figure 6. Examples of our ML approach correctly identifying very localized point sources around the edge of salt flats in Bolivia and Chile. Notice the narrow dust plumes originating from precisely the identified source regions that have been highlighted in blue and cyan.

Identifying dust sources is a critical yet challenging task for the accurate simulation of atmospheric particulate distributions (Ginoux et al., 2001 and Prospero et al., 2002) relevant to air quality (Schauer et al., 1996) and climate change (Tegen and Lacis, 1996 and Tegen et al., 1996b).

For this case, we take a new and radically different approach to any previous studies that have sought to identify global dust sources on a routine basis. We demonstrate that this new approach employing ML is very effective. The approach uses *multi-wavelength spectral reflectivity signatures* to characterize land surfaces, *naturally paving the way for a new class of algorithm ideally suited to fully exploit the next generation of hyper-spectral instrument*. A common application of remote sensing is production of thematic maps using an image classification (Foody, 2002). New in our approach is that we can both operate at very *high spatial resolution* and *distinguish between types of dust sources*. For example, we can easily distinguish between the edge of salt flats (Fig. 6), dried up wadis or lakes, and agricultural sources to name just three of many examples. The only limiting factor for the resolution is the resolution of the satellite imagery.

We employ ML to *objectively* provide an *unsupervised multivariate and nonlinear classification* into a very large number of surface types (in our demonstration study presented below 1000 classes are used) using multi-spectral satellite data. In other words, we *do not impose any a priori assumptions*, but rather, we *let the data speak for itself* as to how we should classify surface types. Self-organizing maps (SOMs) are good candidates to handle this classification task. SOMs are a data visualization and*unsupervised classification technique* invented by Kohonen (1982). They reduce the dimensions of data through the use of self-organizing neural networks. SOMs help us address the issue that humans simply cannot visualize high dimensional data unaided. The way SOMs go about reducing dimensionality is by producing a feature *map*, usually with two dimensions, that objectively plots the *similarities* of the data by *grouping similar data items together*. SOMs learn to classify input vectors according to how they are grouped in the input space. The SOM learns to recognize neighboring sections of the input space. Thus, SOMs learn both the distribution and topology of the input vectors they are trained on. This approach allows SOMs to display similarities and reduce the dimensionality. A SOM does not assume a priori a functional form for the analyzed data.*A noteworthy enhancement of an SOM over principal component analysis is an SOM's ability to represent non-linear functions or mappings*. For further details please refer toKohonen (1982).

The premise being that there are very many types of dust sources, from the diatom rich sediments of the Bodélé depression in Chad, to those at the edge of salt flats in Bolivia and Chile (Fig. 5), to those in the coastal Green Mountains of Libya. Each of these dust sources has *distinct physical characteristics*, and therefore a *distinct reflectance signature*. If we are able to identify these signatures, then we can map the temporal and spatial evolution of each of these distinct dust sources. Once we have the surface type classification, we then seek to identify which small subset of surface classes correspond to various kinds of dust sources. Once we have identified the signature of a wide variety of dust sources, we can precisely pick out these locations globally and how their distribution changes with time. This is particularly useful as dust sources are very localized, and some dust sources have a significant seasonal time evolution. Having a methodology to identify the signature of these small-scale regions is invaluable.

The ML approach to dust source identification was first conceived in 2010 to face a very practical challenge that the Navy has in producing real time visibility forecasts (Walker et al., 2009). If the standard type of dust sources are used (Ginoux et al., 2001), very poor regional visibility forecasts result. However, the quality of the Navy visibility forecasts drastically improved with an analyst (Annette Walker) manually identifying individual dust sources at the heads of plumes by examining sequences of satellite images such as those shown in Fig. 5 and also the EUMETSAT RGB Composites Dust images available online (http://oiswww.eumetsat.org/IPPS/html/MSG/RGB/DUST/). This methodology is very labor intensive and does not lend itself to easy automation. The first prototype dust sources using the ML approach described here were devised specifically to automate the dust source identification and also allow for the accurate diagnosis of the time evolution in the spatial extent of the dust sources.

Beyond the applications of accurate dust sources for visibility and air quality forecasts, the radiative forcing (RF) due to dust is a key concept in *climate change* calculations considered by the IPCC for the quantitative comparison of the strength of different human and natural agents causing climate change (Solomon et al., 2007). Radiative forcing can be categorized into direct and indirect effects. A significant part of the direct effect is the mechanism by which aerosols scatter and absorb shortwave and longwave radiation, thereby altering the radiative balance of the Earth–atmosphere system.*Mineral dust is a major component of global aerosols that exert a significant direct radiative forcing*. Mineral dust aerosols are produced both naturally (≈70%) and anthropogenically (≈30%).

As discussed above, the main goal is to determine all dust source surface locations on the planet. To this aim, SOM was used to classify all the land surface locations into a very large set of n categories. In the examples shown here, $n = 1000$. A small subset of these 1000 categories will be regions that are dust sources. Naturally, there are a variety of distinct types of dust sources (e.g. dry river beds, agricultural sources, edge of salt flats, etc.) that we would like to delineate.

To achieve a comprehensive classification, we want to consider the conditions present throughout the year, therefore, in the demonstration, we took an entire year of the 0.05° MCD43C3 data product. For this entire year of data, we then calculate the mean, μ for each grid point. This is a massive dataset, and the computational time and memory required to perform the SOM classification increases with the number of data records. For the examples shown here, we therefore first restricted our attention to those broad MODIS surface types that may include dust sources, namely: barren or sparsely vegetated surfaces, croplands, grasslands, and open and closed shrublands. These are MODIS surface types 16, 12, 10, 7 and 6, respectively. For each of these surface types, we then constructed an input vector that contains 7 values, namely for each of the 7 bands provided in the MCD43C3 MODIS product the mean, μ, of the directional and bihemispherical reflectance. When training the SOM algorithm, we used the Euclidean distance to compare the input vectors (each containing 7 values).

In order to provide a fine gradation of classification, we used the SOM to group together the surface locations into 1000 classes, only a small subset of which correspond to regions that are dust sources. Once the classes that correspond to dust sources have been successfully identified, we have an automated method with which we can identify dust sources that can be routinely executed to provide a regular dust source data product. We utilized the extensive hand classification of very localized dust sources produced by the Navy (Walker et al., 2009) for the Middle East and South West Asia to guide our initial determination of which of the 1000 classes are dust sources. It is worth noting that the SOM classes are unique and distinct.

Bolivia and Chile salt flats dust event

Let us examine a case study. Fig. 6 shows the dust event of July 18, 2010 in the Bolivian Altiplano. This event can be seen clearly in the MODIS Aqua True Color image where dust plumes emanate from fluvio-lacustine deposits and fluviodeltaic sediments (Risacher and Fritz, 1991a and Risacher and Fritz, 1991b) around the Salars de Coipasa and Uyuni, Lake Poopo, and other smaller salt flats and lakes. Overlaid are the SOM classes that coincide with active dust sources on the Altiplano. Notice that *the salt flats themselves are not dust sources, rather we see the plumes forming around the edges of the flats and lakes*. SOMs are very successful in identifying the unique spectral signatures of dust sources. A set of papers is in preparation will be describing an exhaustive atlas of the global dust sources.

CONCLUSION AND FUTURE DIRECTION

We have discussed the main areas where ML can make a major impact in geosciences and remote sensing. ML focuses on the automatically extraction of information from data by computational and statistical methods. Herein, the features of the ML techniques for nonparametric regression and classification purposes are outlined. The ML's application areas are very diverse and include different themes such as trace gases, aerosol products, vegetation indices, ocean products, characterization of rock mass, liquefaction phenomenon, ground motion parameters, interpreting the remote sensing image, etc. We also presented a review of a number of recent applications of the new GP method in the field. Two illustrative examples are presented to demonstrate the efficiency of ML for tackling the geosciences and remote sensing problems. Currently, data analysis methods play a central role in geosciences and remote sensing. While gathering large collections of data is essential in the field, analyzing this information becomes more challenging. Evidently, such “Big Data” has notable effects both on the predictive analytics and the knowledge extraction and interpretation tools. Considering the significant capabilities of ML, it seems to be a very efficacious approach to handle this type of information.

REFERENCES

1. Alavi, A.H., Gandomi, A.H., 2012. Energy-based models for assessment of soil liquefaction. Geoscience Frontiers 3 (4), 541e555.
2. Alavi, A.H., Gandomi, A.H., Modaresnezhad, M., Mousavi, M., 2011b. New groundmotion prediction equations using multi expression programming. Journal of Earthquake Engineering 15 (4), 511e536.
3. Alavi, A.H., Gandomi, A.H., 2011. A robust data mining approach for formulation of geotechnical engineering systems. Engineering Computations 28 (3), 242e274.
4. Alavi, A.H., Ameri, M., Gandomi, A.H., Mirzahosseini, M.R., 2011a. Formulation of flow number of asphalt mixes using a hybrid computational method. Construction and Building Materials 25 (3), 1338e1355.
5. Alavi, A.H., Gandomi, A.H., Sahab, M.G., Gandomi, M., 2010. Multi expression programming: a new approach to formulation of soil classification. Engineering with Computers 26 (2), 111e118.
6. Allen, M.R., Stott, P.A., Mitchell, J.F.B., Schnur, R., Delworth, T.L., 2000. Quantifying the uncertainty in forecasts of anthropogenic climate change. Nature 407 (6804), 617e620.
7. Atkinson, P.M., Tatnall, A.R.L., 1997. Introduction: neural networks in remote sensing. International Journal of Remote Sensing 18 (4), 699e709.
8. Ayala, A., Brauer, M., Mauderly, J.L., Samet, J.M., 2012. Air pollutants and sources associated with health effects. Air Quality Atmosphere and Health 5 (2), 151e167. http://dx.doi.org/10.1007/s11869-011-0155-2.
9. Azamathulla, H.M., Ghani, A.A., Fei, S.Y., 2012. ANFIS-based approach for predicting sediment transport in clean sewer. Applied Soft Computing 12 (3), 1227e1230.
10. Azamathulla, H.M., Guven, A., Demir, Y.K., 2011. Linear genetic programming to scour below submerged pipeline. Ocean Engineering 38 (8e9), 995e1000.
11. Azamathulla, H.M., Wu, F.C., 2011. Support vector machine approach for longitudinal dispersion coefficients in natural streams. Applied Soft Computing 11 (2), 2902e2905.
12. Azamathulla, H.M., 2012. Linear programming for irrigation scheduling e a case study (Book Chapter). In: Linear Programming: New Frontiers in Theory and Applications, pp. 174e192.
13. Baykasoglu, A., Gullu, H., Canakci, H., Ozbakir, L., 2008. Prediction of compressive and tensile strength of limestone via genetic programming. Expert Systems with Applications 35 (1e2), 111e123.

14. Beiki, M., Bashari, A., Majdi, A., 2010. Genetic programming approach for estimating the deformation modulus of rock mass using sensitivity analysis by neural network. International Journal of Rock Mechanics and Mining Sciences 47, 1091e1103.
15. Boldo, E., Linares, C., Lumbreras, J., Borge, R., Narros, A., Garcia-Perez, J., FernandezNavarro, P., Perez-Gomez, B., Aragones, N., Ramis, R., Pollan, M., Moreno, T., Karanasiou, A., Lopez-Abente, G., 2011. Health impact assessment of a reduction in ambient PM2.5 levels in spain. Environment International 37 (2), 342e348. http://dx.doi.org/10.1016/j.envint.2010.10.004.
16. Brook, R.D., Rajagopalan, S., Pope, I., Arden, C., Brook, J.R., Bhatnagar, A., DiezRoux, A.V., Holguin, F., Hong, Y., Luepker, R.V., Mittleman, M.A., Peters, A., Siscovick, D., Smith, J., Sidney, C., Whitsel, L., Kaufman, J.D., E. Amer Heart Assoc Council, D. Council Kidney Cardiovasc, M. Council Nutr Phys Activity, 2010.
17. Particulate matter air pollution and cardiovascular disease an update to the scientific statement from the American heart association. Circulation 121 (21), 2331e2378. http://dx.doi.org/10.1161/CIR.0b013e3181dbece1.
18. Brown, M.E., Lary, D.J., Vrieling, A., Stathakis, D., Mussa, H., 2008. Neural networks as a tool for constructing continuous NDVI time series from AVHRR and MODIS. International Journal of Remote Sensing 29 (24), 7141e7158.
19. Cabalar, A.F., Cevik, A., 2009. Genetic programming-based attenuation relationship, an application of recent earthquakes in turkey. Computers & Geosciences 35 (9), 1884e1896.
20. Carpenter, G.A., Gjaja, M.N., Gopal, S., Woodcock, C.E., 1997. Art neural networks for remote sensing: vegetation classification from landsat tm and terrain data. IEEE Transactions on Geoscience and Remote Sensing 35 (2), 308e325.
21. Chen, L., Yeh, K., Wei, H., Liu, G., 2011. An improved genetic programming to SSM/I estimation typhoon precipitation over ocean. Hydrol Process 25, 2573e2583.
22. Chen, L., 2003. A study of applying genetic programming to reservoir trophic state evaluation using remote sensor data. International Journal of Remote Sensing 24 (11), 2265e2275.
23. Chuang, M.-T., Zhang, Y., Kang, D., 2012. Application of wrf/chem-madrid for realtime air quality forecasting over the southeastern united states (vol 45, pg 6241, 2011). Atmospheric Environment 60, 677e678.
24. Das, S.K., Basudhar, P.K., 2008. Prediction of residual friction angle of clays using artifical neural network. Engineering Geology 100 (3e4), 142e145.

25. Dockery, D.W., Pope, C.A., Xu, X.P., Spengler, J.D., Ware, J.H., Fay, M.E., Ferris, B.G., Speizer, F.E., 1993. An association between air-pollution and mortality in 6 united-states cities. New England Journal of Medicine 329 (24), 1753e1759. http://dx.doi.org/10.1056/nejm199312093292401. //WOS: A1993MK09600001.
26. dos Santos, J.A., Faria, F.A., Calumby, R.T., Torres, R., da, S., Lamparelli, R.A.C., July 2010. A genetic programming approach for coffee crop recognition. In: IGARSS 2010. Honolulu, USA, pp. 3418e3421.
27. Engel-Cox, J.A., Hoff, R.M., Haymet, A.D.J., 2004a. Recommendations on the use of satellite remote-sensing data for urban air quality. Journal of the Air & Waste Management Association 54 (11).
28. Engel-Cox, J.A., Holloman, C.H., Coutant, B.W., Hoff, R.M., 2004b. Qualitative and quantitative evaluation of MODIS satellite sensor data for regional and urban scale air quality. Atmospheric Environment 38 (16).
29. Foody, G.M., 2002. Status of land cover classification accuracy assessment. Remote Sensing of Environment 80 (1), 185e201. http://dx.doi.org/10.1016/S0034- 4257(01)00295-4.
30. Gandomi, A.H., Alavi, A.H., Mousavi, M., Tabatabaei, S.M., 2011. A hybrid computational approach to derive new ground-motion prediction equations. Engineering Applications of Artificial Intelligence 24 (4), 717e732.
31. Gandomi, A.H., Alavi, A.H., 2013. Hybridizing genetic programming with orthogonal least squares for modeling of soil liquefaction. International Journal of Earthquake Engineering and Hazard Mitigation 1 (1), 1e8.
32. Gandomi, A.H., Alavi, A.H., 2011. Multi-stage genetic programming: a new strategy to nonlinear system modeling. Information Sciences 181 (23), 5227e5239.
33. Garg, A., Garg, Ankit, Tai, K., Sreedeep, S., 2014a. An integrated SRM-multi-gene genetic programming approach for prediction of factor of safety of 3-D soil nailed slopes. Engineering Applications of Artificial Intelligence 30, 30e40.
34. Garg, A., Tai, K., Savalani, M.M., 2014b. State-of-the-Art in empirical modeling of rapid prototyping processes. Rapid Prototyping Journal 20 (2), 164e178.
35. Garg, A., Tai, K., Gupta, A.K., 2014c. A modified multi-gene genetic programming approach for modelling true stress of dynamic strain aging regime of austenitic stainless steel 304. Meccanica 49 (5), 1193e1209.
36. Ginoux, P., Chin, M., Tegen, I., Prospero, J.M., Holben, B., Dubovik, O., Lin, S.J., 2001. Sources and distributions of dust aerosols simulated with the GOCART model. Journal of Geophysical Research-Atmospheres 106 (D17), 20255e20273.

37. Ginoux, P., Torres, O., 2003. Empirical TOMS index for dust aerosol: applications to model validation and source characterization. Journal of Geophysical ResearchAtmospheres 108 (D17). http://dx.doi.org/10.1029/2003jd003470.
38. Grell, G.A., Peckham, S.E., Schmitz, R., McKeen, S.A., Frost, G., Skamarock, W.C., Eder, B., 2005. Fully coupled "online" chemistry within the WRF model. Atmospheric Environment 39 (37), 6957e6975.
39. Hansen, J., Fung, I., Lacis, A., Rind, D., Lebedeff, S., Ruedy, R., Russell, G., Stone, P., 1988. Global climate changes as forecast by Goddard institute for space studies 3-dimensional model. Journal of Geophysical Research-Atmospheres 93 (D8), 9341e9364.
40. IPCC, 2013. Climate Change 2013: The Physical Science Basis. Contribution of Working Group I to the Fifth Assessment Report of the Intergovernmental Panel on Climate Change. Cambridge University Press, Cambridge. Javadi, A.A., Rezania, M., Nezhad, M.M., 2006. Evaluation of liquefaction induced lateral displacements using genetic programming. Computers and Geotechnics 33 (4e5), 222e233.
41. Karakus, M., 2011. Function identification for the intrinsic strength and elastic properties of granitic rocks via genetic programming (GP). Computers & Geosciences 37, 1318e1323.
42. Kohonen, T., 1982. Self-organized formation of topologically correct feature maps. Biological Cybernetics 43 (1), 59e69.
43. Koza, J., 1992. Genetic Programming, on the Programming of Computers by Means of Natural Selection. MIT Press, Cambridge (MA).
44. Lary, D.J., Remer, L.A., MacNeill, D., Roscoe, B., Paradise, S., 2009. Machine learning and bias correction of MODIS aerosol optical depth. IEEE Geoscience and Remote Sensing Letters 6 (4), 694e698.
45. Lary, D.J., Muller, M.D., Mussa, H.Y., 2004. Using neural networks to describe tracer correlations. Atmospheric Chemistry and Physics 4, 143e146.
46. Lary, D.J., Faruque, F., Malakar, N., Moore, A., Roscoe, B., Adams, Z., Eggelston, Y., 2014a. Estimating the global abundance of ground level presence of microscopic particulate matter. Geospatial Health 8, S611eS630.
47. Lary, D.J., 2010. Artificial intelligence in geoscience and remote sensing. In: Imperatore, P., Riccio, D. (Eds.), Geoscience and Remote Sensing, New Achievements. IN-TECH, Vukovar, Croatia, pp. 1e24.
48. Lary, D.J., Waugh, D.W., Douglass, A.R., Stolarski, R.S., Newman, P.A., Mussa, H., 2007.
49. Variations in stratospheric inorganic chlorine between 1991 and 2006. Geophysical Research Letters 34.

50. Lee, H.J., Liu, Y., Coull, B., Schwartz, J., Koutrakis, P., 2011a. PM2.5 prediction modeling using MODIS AOD and its implications for health effect studies. Epidemiology 22 (1).
51. Lee, H.J., Liu, Y., Coull, B.A., Schwartz, J., Koutrakis, P., 2011b. A novel calibration approach of MODIS AOD data to predict PM2.5 concentrations. Atmospheric Chemistry and Physics 11 (15).
52. Lewkowski, C., Porwal, A., González-Álvarez, I., 2010. Genetic programming applied to base-metal prospectivity mapping in the Aravalli Province, India. Geophysical Research Abstracts 12, EGU2010e15171.
53. Lia, W.X., Daib, L.F., Houa, X.B., Leia, W., 2007. Fuzzy genetic programming method for analysis of ground movements due to underground mining. International Journal of Rock Mechanics and Mining Sciences 44, 954e961.
54. Liu, Y., Franklin, M., Kahn, R., Koutrakis, P., 2007. Using aerosol optical thickness to predict ground-level PM2.5 concentrations in the St. Louis area: a comparison between MISR and MODIS. Remote Sensing of Environment 107 (1e2).
55. Liu, Y., He, K., Li, S., Wang, Z., Christiani, D.C., Koutrakis, P., 2012. A statistical model to evaluate the effectiveness of PM2.5 emissions control during the beijing 2008 olympic games. Environment International 44.
56. Liu, Y.-J., Harrison, R.M., 2011. Properties of coarse particles in the atmosphere of the united kingdom. Atmospheric Environment 45 (19).
57. Madadi, M.R., Azamathulla, H. Md, Yakhkeshi, M., 2015. Application of Google Earth to investigate the change of flood inundation area due to flood detention dam. Earth Science Informatics. http://dx.doi.org/10.1007/s12145-014-0197-8.
58. Makkeasorn, A., Chang, N.B., Li, J., 2009. Seasonal change detection of riparian zones with remote sensing images and genetic programming in a semi-arid watershed. Journal of Environmental Management 90, 1069e1080.
59. Makkeasorn, A., Chang, N.B., Beaman, M., Wyatt, C., Slater, C., 2006. Soil moisture estimation in a semi-arid watershed using RADARSAT-1 satellite imagery and genetic programming. Water Resources Research 42, 1e15.
60. Nikravesh, M., 2007. Computational intelligence for geosciences and oil exploration. In: Forging New Frontiers: Fuzzy Pioneers I, vol. 66. California University Press, pp. 267e332.
61. Ozbek, A., Unsal, M., Dikec, A., 2013. Estimating uniaxial compressive strength of rocks using GEP. Journal of Rock Mechanics and Geotechnical Engineering 5, 325e329.
62. Pope, C., Dockery, A.,D.W., 2006. Health effects of fine particulate air pollution: lines that connect. Journal of the Air & Waste Management Association 56 (6), 709e742.

63. Pope, I., Arden, C., Burnett, R.T., Krewski, D., Jerrett, M., Shi, Y., Calle, E.E., Thun, M.J., 2009. Cardiovascular mortality and exposure to airborne fine particulate matter and cigarette smoke shape of the exposure-response relationship. Circulation 120 (11), 941e948. http://dx.doi.org/10.1161/circulationaha.109.857888.
64. Prospero, J.M., Ginoux, P., Torres, O., Nicholson, S.E., Gill, T.E., 2002. Environmental characterization of global sources of atmospheric soil dust identified with the nimbus 7 total ozone mapping spectrometer (TOMS) absorbing aerosol product. Reviews of Geophysics 40 (1), 31.
65. Prospero, J.M., 2003. Global dust transport over the oceans: the link to climate. Geochimica Et Cosmochimica Acta 67 (18). A384eA384.
66. Rajeev, K., Parameswaran, K., Nair, S.K., Meenu, S., 2008. Observational evidence for the radiative impact of Indonesian smoke in modulating the sea surface temperature of the equatorial Indian Ocean. Journal of Geophysical Research-Atmospheres 113 (D17).
67. Ravandi, E.G., Rahmannejad, R., Feili Monfared, A.E., Ravandid, E.G., September 2013. Application of numerical modeling and genetic programming to estimate rock mass modulus of deformation. International Journal of Mining Science and Technology 23 (5), 733e737.
68. Risacher, F., Fritz, B., 1991a. Geochemistry of bolivian salars, lipez, southern altiplano origin of solutes and brine evolution. Geochimica Et Cosmochimica Acta 55 (3), 687e705.
69. Risacher, F., Fritz, B., 1991b. Quaternary geochemical evolution of the salars of uyuni and coipasa, central altiplano, bolivia. Chemical Geology 90 (3e4), 211e231.
70. Rosin, P., Hervas, J., 2002. Image thresholding for landslide detection by genetic programming. In: Bruzzone, L., Smiths, P. (Eds.), Analysis of Multi-temporal Remote Sensing Images. World Scientific, pp. 65e72.
71. Samui, P., 2008a. Support vector machine applied to settlement of shallow foundations on cohesionless soils. Computers and Geotechnics 35 (3), 419e427.
72. Samui, P., 2008b. Slope stability analysis: a support vector machine approach. Environmental Geology 56 (2), 255e267, 35.
73. Samui, P., 2012. Application of statistical learning algorithms to ultimate bearing capacity of shallow foundation on cohesionless soil. International Journal for Numerical and Analytical Methods in Geomechanics 36 (1), 100e110.
74. Schaap, M., Apituley, A., Timmermans, R.M.A., Koelemeijer, R.B.A., de Leeuw, G., 2009. Exploring the relation between aerosol optical depth and PM2.5 at Cabauw, the Netherlands. Atmospheric Chemistry and Physics 9 (3).

75. Schauer, J., Rogge, W., Hildemann, L., Mazurek, M., Cass, G., Simoneit, B., 1996.
76. Source apportionment of airborne particulate matter using organic compounds as tracers. Atmospheric Environment 30 (22), 3837e3855.
77. Shahin, M.A., Jaksa, M.B., 2005. Neural network prediction of pullout capacity of marquee ground anchors. Computers and Geotechnics 32 (3), 153e163.
78. Shahin, M.A., Jaksa, M.B., Maier, H.R., 2001. Artificial neural network applications in geotechnical engineering. Australian Geomechanics 36 (1), 49e62.
79. Shuhua, Z., Qian, G., Jianguo, S., 2006. Genetic programming approach for predicting surface subsidence induced by mining. Journal of China University of Geosciences 17 (4), 361e366.
80. Solomon, S., Qin, D., Manning, M., Chen, Z., Marquis, M., Averyt, K., Tignor, M., Miller, H., 2007. IPCC Fourth Assessment Report: Climate Change 2007. Cambridge University Press, Cambridge.
81. Tegen, I., Lacis, A., Fung, I., 1996. The influence of mineral aerosols from disturbed soils on the global radiation budget. Nature 380, 419e422.
82. Tegen, I., Lacis, A., 1996. Modeling of particle size distribution and its influence on the radiative properties of mineral dust aerosol. Journal of Geophysical Research 101, 19237e19244.
83. Walker, A.L., Liu, M., Miller, S.D., Richardson, K.A., Westphal, D.L., 2009. Development of a dust source database for mesoscale forecasting in southwest Asia. Journal of Geophysical Research-Atmospheres 114.
84. WHO, 2014. 7 million Premature Deaths Annually Linked to Air Pollution. URL: http://www.who.int/mediacentre/news/releases/2014/air-pollution/en/.
85. Yi, J.S., Prybutok, V.R., 1996. A neural network model forecasting for prediction of daily maximum ozone concentration in an industrialized urban area. Environmental Pollution 92 (3), 349e357.
86. Zahabiyoun, B., Goodarzi, M.R., Bavani, A.R.M., Azamathulla, H.M., 2013. Assessment of climate change impact on the Gharesou river Basin using SWAT Hydrological model. Clean e Soil, Air, Water 41 (6), 601e609.

CITATION

David J. Lary, Amir H. Alavi, Amir H. Gandomi, Annette L. Walker, Machine learning in geosciences and remote sensing, Geoscience Frontiers, Available online 12 August 2015, ISSN 1674-9871, http://dx.doi.org/10.1016/j.gsf.2015.07.003.

CHAPTER 5

Geophysical, Remote Sensing, Gis, and Isotopic Applications for a Better Understanding of the Structural Controls on Groundwater Flow in the Mojave Desert, California

D. Dailey[1], W. Sauck[1], M. Sultan[1], A. Milewski[2], M. Ahmed[1,3], W.R. Laton[4], R. Elkadiri[1], J. Foster[4], C. Schmidt[1], T. Al Harbi[1,5]

[1] Department of Geosciences, Western Michigan University, Kalamazoo, MI 49008, USA

[2] Department of Geology, University of Georgia, Athens, GA 30602, USA

[3] Department of Geology, Faculty of Science, Suez Canal University, Ismailia 41522, Egypt

[4] Department of Geology, California State University Fullerton, Fullerton, CA 92834, USA

[5] Department of Geology and Geophysics, King Saud University, Riyadh 11451, Saudi Arabia

HIGHLIGHTS

- Faults and uplifts control on groundwater flow in Mojave Desert was investigated.
- Shallow basement parallel to and west of Helendale Fault (HF) was detected (Vertical Electrical Soundings [VESs]).
- Conductive water-saturated breccia was detected (Very Low Frequency [VLF]) along HF.
- HF channels groundwater along strike; uplifts restrict lateral flow/infiltration.
- Findings invite investigations in similar settings in Mojave and elsewhere worldwide.

ABSTRACT

Study region

Study focus

An integrated (near-surface geophysics, remote sensing, isotopic analyses) study was conducted in the Mojave River Basin and Morongo Groundwater Basin to investigate potential effects that the Helendale Fault [HF] and basement uplifts might have on groundwater flow in the Mojave Desert.

New hydrological insights for the region

The HF traces were mapped using LiDAR and Geoeye-1 imagery (surface) and magnetic profiles (subsurface). Shallow basement parallel to and west of the HF was detected using the Vertical Electrical Soundings (VESs). Conductive water-saturated breccia was detected along the HF using the Very Low Frequency (VLF) electromagnetic measurements. Isotopic analyses (δD and $\delta^{18}O$) for groundwater samples from productive shallow wells, and springs sampled west of the HF and the basement uplift are less depleted (Group I: Fifteenmile Valley Groundwater sub-basin [FVGS]; average δD: −86.8‰; $\delta^{18}O$: −11.8‰) than samples east of the basement uplift (Group II: Lucerne Valley Groundwater sub-basin [LVGS]; average δD: −95.0‰; $\delta^{18}O$: −12.1‰), whereas samples proximal to, the fault have compositions similar to Group I but show evidence for mixing with Group II compositions (Group III; average δD: −88.8‰; $\delta^{18}O$: −11.5‰). Findings are consistent with the HF channeling groundwater from the San Bernardino Mountains with basement uplifts acting as barriers to lateral groundwater flow and could be applicable to similar settings across the Mojave Desert and elsewhere worldwide.

INTRODUCTION

The structural controls on groundwater accumulation and flow exhibit a wide range of variability. Faults can act as highly permeable pathways, or as barriers for groundwater flow depending on the aquifer lithology, aquifer hydrologic conditions, and fault characteristics (Bense and Person, 2006, Celico et al., 2006 and Dewandel et al., 2006). The extent to which groundwater flows along the strike of a fault is controlled by fault properties (e.g., width of core and deformation zones, parent material), hydrologic framework (e.g., permeability of the individual fault rocks/fractures, geometric architecture in three dimensions, extent of clay core), and hydrologic parameters (e.g., fault specific resistance, transmissibility, hydraulic head) (Bense et al., 2003, Lunn et al., 2008, Caine and Minor, 2009 and Faulkner et al., 2010). Various mechanisms were suggested to explain the damming effect of the fault zones across unconsolidated sediments including the cementation or mineralization of the fault zone, grain crushing and grain realignment and juxtaposition seals (Bense and Person,

2006 and Knipe, 1993). In this study we adopt an integrated (geophysics, remote sensing, GIS, and isotopic analyses) approach to investigate the potential role of structural elements, faults and basement uplifts, in controlling the groundwater flow in the Mojave Desert.

The Mojave Desert occupies areas in southeastern and central California, southern Nevada, southwestern Utah, and northwestern Arizona in the United

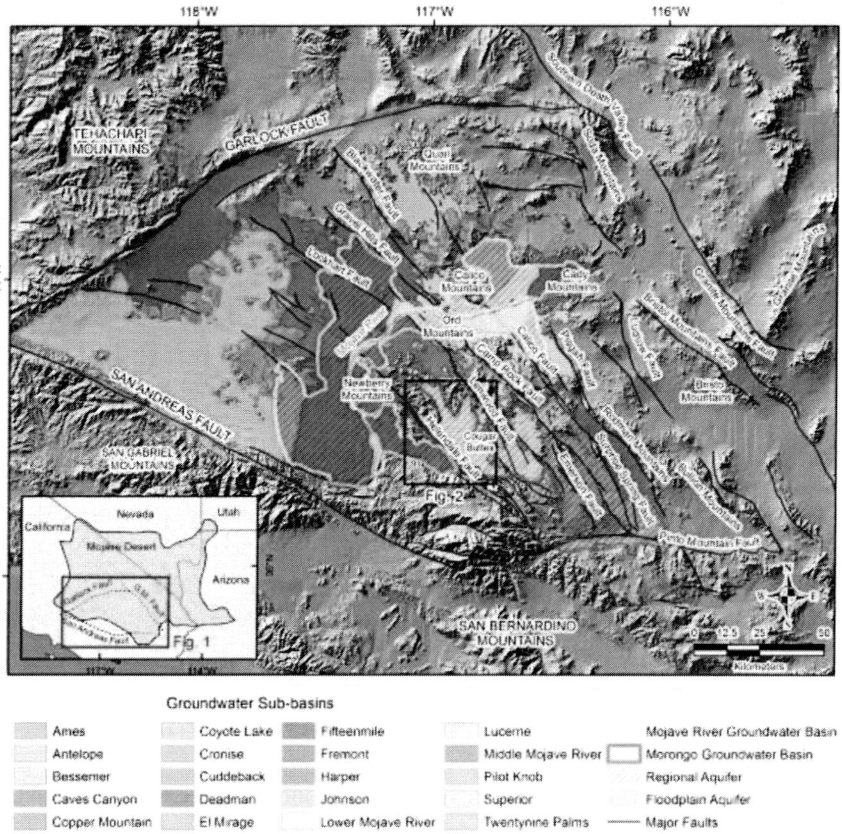

Figure 1. Overview of the southern Mojave Desert, with hillshade applied to accentuate topographic features showing the distribution of major (1) faults (e.g., Helendale, San Andreas); (2) mountain ranges (e.g., Tehachapi, San Bernardino); (3) groundwater basins (e.g., Mojave River, Morongo); (4) groundwater sub-basins (e.g., Lucerne Valley, Fifteenmile Valley); and (5) aquifers (e.g., Regional, Floodplain). Enlargements of the study area are shown in Fig. 2A and B. The inset (lower left corner) shows the areal extent of the Mojave Desert in California, Nevada, Utah, and Arizona. Areas covered by Fig. 1 and Fig. 2are outlined by boxes in the inset and in Fig. 1, respectively.

States (Fig. 1 inset). The rapidly growing population (1980 population: 64,685; 2010 population: 349,730), low humidity, high summer temperature, and the paucity of precipitation (100–140 mm/year; Londquist and Martin, 1991) in the Mojave Desert has created an ever-increasing need for freshwater resources. Surface water is limited to ephemeral flow during winter and spring storm periods.

Two main groundwater basins were recognized in the southwestern parts of the Mojave Desert, the Mojave River Groundwater Basin (MRGB), and the Morongo Groundwater Basin (MGB) (Fig. 1). Runoff and groundwater flow from the melting of snowpack over the adjacent San Bernardino Mountains constitute the principal sources of groundwater in the southern parts of the MRGB and MGB (Stamos et al., 2003 and USGS, 2004), whereas the northern and central portions of the basin are largely recharged by groundwater and surface flow from surrounding mountain ranges such as the Cougar Buttes and Granite Mountains (Fig. 1), as well as infiltration via the Mojave River floodplain aquifer.

Increasing extraction of groundwater from the MRGB and the MRB has affected various sections in the basin to varying degrees. The central portions of the MRGB witnessed a progressive decline in water levels over the years (20 m from 1950 to 1986), whereas areas proximal to the mountains showed minimal or significantly less decline (Smith and Pimentel, 1998). One popular explanation for the groundwater deficits in the central parts of the basin is that the dextral faults in the area acted as barriers for lateral (across-fault) flow of groundwater from the mountains toward the central parts of the basin. A better understanding of how regional and local fault systems interact with the principle aquifers in the Mojave Desert is of key importance in meeting the rising demand for groundwater resources.

In this study, we examine the role of one of the major dextral faults in the MRGB and the MRB, the Helendale Fault (HF), and its subsequent splays as potential barriers (across strike) to groundwater flow or as enhanced groundwater flow pathways along its strike that probably feed both basins. The former widely accepted hypothesis is largely based on the observed water level differences (~40 m) across the HF in the Central Lucerne Valley area and those (3–9 m) across both the Johnson Valley Fault and the Emerson Fault (Lewis, 1972, Trayler and Koczot, 1995 and GSI, 2000). This explanation is at odds with reported findings elsewhere that suggest that fractured fault planes often act as enhanced groundwater flow directions along strike (e.g., Barton et al., 1995, Caine et al., 1996, Gudmundsson, 2001 and Sultan et al., 2007). If such models were applicable to the HF, it could be channeling groundwater from the San Bernardino Mountains in the south to the lowlands of the MRGB and MGB in the north.

An integrated approach utilizing remote sensing (LiDAR and GEOEye-1 images), geophysical (Very Low Frequency [VLF], magnetic, and Vertical Electrical Soundings [VES]) data, isotopic (O, H stable isotope) compositions for groundwater samples, and subsurface well data was adopted to investigate (1) the potential role of the HF as a barrier for lateral groundwater flow or as an enhanced groundwater flow pathway along its strike, and (2) the distribution and nature of barriers for lateral groundwater flow in the study area. Specifically, we investigate a previously unrecognized preferential groundwater flow path from mountainous source areas that could be providing additional contributions to the aquifers in the study area. Specifically, we investigate the potential role of the strike-slip faults in the basin as enhanced groundwater flow pathways along strike that could provide recharge to both local and regional aquifers. We also investigate the previously-unrecognized role of basement uplifts (ridges) in controlling lateral groundwater flow in the study area. The wide distribution of similar structural elements (e.g., strike-slip faults) and landforms (mountainous areas and inland basins) throughout the MRGB, MGB, and large sectors of the Mojave Desert makes it likely that our findings could be applicable to many other similar surrounding areas.

GEOLOGIC AND HYDROGEOLOGIC SETTING

The southwestern Mojave Desert is characterized by the numerous mountain ranges within it, including the Tehachapi, San Bernardino, Bullion, Quail, Soda, and Calico ranges, which are separated by valleys and inland basins (Fig. 1). These mountain ranges and valleys are part of the Basin and Range province, an area undergoing crustal thinning, pull-apart basin tectonics, and lateral displacements across the numerous dextral faults (e.g., Helendale, Lenwood, Lockhart, and Camp Rock) related to the Late Cenozoic tectonics in the area (Dokka, 1983 and Dokka and Travis, 1990). These dextral fault systems formed secondarily to the San Andreas Fault, primarily as a result of extension and rotation of the Mojave Fault Block (Dokka and Travis, 1990 and Dokka and Macaluso, 2001).

The MRGB (area: 3630 km^2) is bounded by the San Bernardino and San Gabriel Mountains in the south, by the Lucerne Valley to the east, and by the Antelope Valley to the west (Fig. 1). The MGB (area: 2600 km^2) is surrounded by the Ord and Granite mountains to the north, the Bullion Mountains to the east, the San Bernardino Mountains to the southwest, and the Little San Bernardino Mountains to the south (Fig. 1). Bedrock units throughout the MRGB and MGB consist mainly of Mesozoic granites and gabbros, with sparse dispersions of Tertiary volcanic units in the central and northern portions of the basin. Quaternary alluvial sediments make up the majority of surface sediments, with additional marine and non-marine sand deposits spread sparsely throughout the basin, generally near the desert playas (Fig. 2A).

Both the MRGB and MGB are characterized by a series of Cenozoic, parallel, northwest-trending dextral strike-slip faults that make up the Mojave Desert Fault Block in the eastern California Shear Zone. Their occurrence spans from the San Andreas Fault in the west to the Johnson Valley and Granite Mountains faults in the east (Fig. 1). The fault system is commonly thought of as a simple shear system with dextral faults terminating against the cross-cutting Garlock Fault to the north and the San Bernardino thrust belt in the south (Garfunkel, 1974). This model is contested by more recent research that suggests a more complex rotational system, with the majority of offset along the faults being attributable to extensional and rotational elements due to movement along the San Andreas Fault (Dokka, 1986 and Dokka and Travis, 1990). The Northeast Mojave Block, located in the northeastern portion of the Mojave Desert, is a separate rotational block of dextral faults, oriented west to east and trending nearly perpendicular to the majority of the other dextral faults in the shear zone (Dokka et al., 1991).

MRGB and MGB are broken up into a series of smaller sub-basins. These sub-basins are apparently separated by ridges defined by basement and/or bedrock outcrops and shallow bedrock saddles between the exposed ridges that are proximal and parallel to the dextral fault systems in the valley. The faults originate and/or terminate against the northern and southern mountain ranges. One such sub-basin of the MGB is the Lucerne Valley Groundwater sub-basin (LVGS), a land-locked basin with limited external surface water flowing out of it (Meisling, 1984); another is the Fifteenmile Valley Groundwater sub-basin (FVGS), also called the “Upper Mojave River Valley Groundwater sub-basin” within the MRGB (Fig. 1). Fig. 2A and B reveal that the LVGS and the FVGS are separated by a northwest-southeast trending ridge that runs parallel and proximal to the HF (Fig. 1).

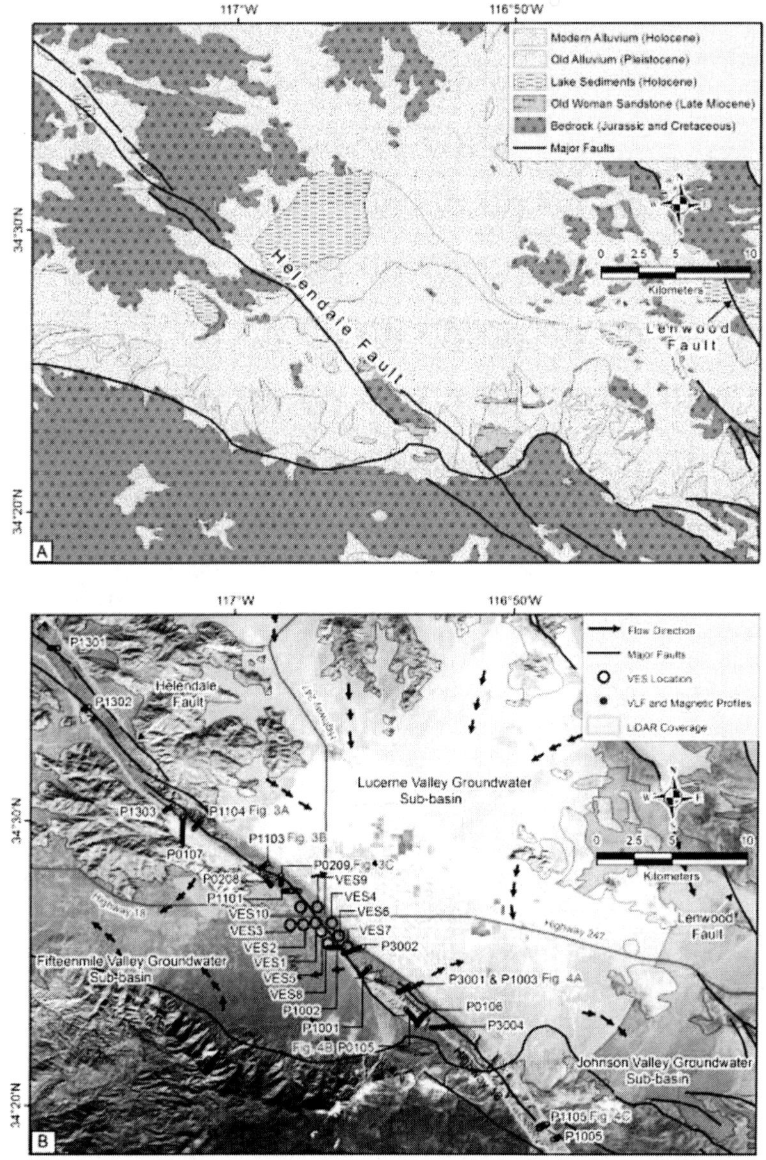

Figure 2. (A) Simplified geologic map (modified from MWA, 2007) showing distribution of the major rock groups and bedrock ridges (subparallel to Helendale and Lenwood faults) that we infer to extend along their projected length at shallow depths beneath the overlying alluvium. (B) Location map for our VLF and magnetic profiles (thick black lines) and VES resistivity data (open circles). Also shown are (1) groundwater sub-basins (i.e., Lucerne Valley, Fifteenmile Valley, and Johnson Valley); and (2) groundwater flow direction within the sub-basins (Stamos and Predmore, 1995, ESA, 2004 and MWA, 2005).

The latter originates in the south along the San Bernardino mountain front and extends northwest across the LVGS, terminating in the Newberry Mountains. Across some of these ridges and proximal to sub-parallel faults, large differences in water levels were reported (Lewis, 1972, MWA, 2007, Trayler and Koczot, 1995 and GSI, 2000). Water levels are not the only features that vary across these topographic ridges and faults. Varying groundwater flow directions were reported from aquifers separated by these topographic highs and faults, as seen in Fig. 2B.

Within the MRGB, an unconsolidated alluvial aquifer, the Floodplain Aquifer, extends along the Mojave River and is surrounded and underlain by an extensive aquifer known as the Regional Aquifer (Fig. 1). These aquifers are the primary sources of groundwater for the residents of Lucerne Valley. The interconnected alluvial basins and topographically closed basins of the Regional Aquifer extend across the entire MRGB and MGB (Izbicki, 2004). The aquifer is formed of unconsolidated older (Pleistocene to Pliocene) alluvium of the ancestral Mojave River and undifferentiated alluvium of Holocene to Pliocene age (Stamos et al., 2001).

METHODOLOGY

We adopted a four-step methodology. First, we collected and processed available satellite and airborne light detection and ranging (LiDAR) images to better delineate the fault traces because many of them are obscured by alluvium or too subtle to observe in the field. Also, the fine-scale fault mapping of all the faults is needed to define the appropriate location for conducting geophysical profiles. Second, we applied several geophysical techniques to verify the satellite-based distribution of faults using magnetic and electromagnetic measurements, and to examine the role of faults and basement uplifts in groundwater flow. Third, we collected groundwater samples from the study area to conduct isotopic analyses to search for isotopic provenances related to the distribution of faults and uplifts in the study area. Last, the collected geophysical and isotopic data were evaluated together with all other relevant hydrologic data (e.g., water levels) in a Geographic Information System (GIS) environment for a better identification of the spatial relationships and understanding of the significance of these relationships.

Remote sensing methods

The airborne Light Detection and Ranging (LiDAR; http://opentopo.sdsc.edu) laser (wavelength: 1047×10^{-9} m) emission and detection system produces detailed digital elevation images (spatial resolution: 1 m) of the surface while stripping away surface obscurers such as vegetation (Haugerud et al., 2003). The available LiDAR segments (12 segments) over and around the HF were used to identify small-scale features, such as subtle topographic changes and partially obscured fault features.

In areas were the LiDAR data are not available, the very high (panchromatic: 0.41 m, multispectral: 1.65 m) spatial resolution Google Earth GeoEye-1 images (www.geoeye.com) were used for examining smaller, less-regional features, including, but not limited to, fault splays. The LiDAR and GeoEye-1 data were used to better map the fault traces and splays of the HF for a better definition of the optimum locations for conducting geophysical surveys.

Geophysical methods

Very Low Frequency (VLF) profiles

The Very Low Frequency (VLF) radio receiver (IRIS T-VLF) was used to measure the distortion of the normally horizontal electromagnetic flux lines by local electrical conductors. The VLF uses the radio carrier waves of the submarine communications stations of the various navies of the world (Paterson and Ronka, 1971). The instrument records each frequency used, the tilt of the electromagnetic field (from the horizontal, given as a percentage), ellipticity in the vertical and horizontal planes, and signal strength.

We primarily used the VLF method (Paterson and Ronka, 1971) in the field to verify the presence of groundwater along the length of the HF and its splays. The VLF is ideal for this application, because it detects conductive water-saturated sub-vertical breccia zones in bedrock (Palacky et al., 1981). If the hypothesis that the dextral faults and intersecting fault splays in the LVGS and the FVGS are channeling groundwater is plausible, a significant conductive response in the VLF would be expected as the profile crosses the fault plane. However, the VLF instrument has several limitations. It responds strongly to massive sulfides, as well as to graphitic shear zones; however, the MRGB and the MGB are not known to contain these types of sulfide deposits and shear zones, so it is unlikely that interference would present itself in this form. It is also limited to detection within an approximately 90° fan of strikes (±45° from the radial azimuth to the transmitting station). To adjust for this, measurements were recorded using one of the three transmitting stations: Seattle, Washington (NLK at 24.8 kHz), which provided nearly ideal geometry for profiles acquired in an east or northeast direction; Lamour, North Dakota (NML at 25.2 kHz), for which only 2011 data was available, was used for the northwest-trending profiles; and Cutler, Maine (NAA at 24.0 kHz), was used as an alternative, since the receiver collects from two stations simultaneously. Data from the Cutler, Maine station was generally only used for north-trending profiles. The VLF method is limited to use during daylight hours, when the overhead ionosphere is well developed (Vallèe et al., 1992). It is also subject to sudden pulsations from the solar wind, which result in rapid deviations of the apparent tilt angle that may last for several minutes. In the event that such disturbances were encountered, data collection was paused or repeated to rule out this possible interference source.

Over the course of three summers (2009–2011), 19 profiles were performed in various locations across or near the HF (Fig. 2B). All profiles were acquired within the same time period (May–June) each year and some of these profiles were measured more than one time during the individual time periods. These profiles were commonly acquired along the flanks of exposed ridgelines along the southern portion of the HF, the lower alluvial portions of the valley, and the canyons in the northern portion of the Lucerne Lake sub-basin. Some of these profiles were measured across suspected splays of the main fault, as opposed to crossing the HF itself. Several of these splays had been mapped previously, while others were identified using available geologic maps and remote-sensing datasets. Seven of these profiles will be detailed in the discussion section of this paper (Section 5).

Magnetic profiles

Magnetic profiles were used to define the subsurface postulated extension of the remote sensing-based (i.e., LiDAR, GeoEye-1) faults and fault splays and to help reveal the location of a fault plane where its location is not readily apparent or well known. Juxtaposition of rock units of varying magnetic susceptibilities usually shows as a magnetic anomaly as the profile crosses the fault plane. The Proton-Precession magnetometer (GeoMetrics G-816) was used to measure the absolute total magnetic field strength (in nano-Teslas [nT]) at the position of a sensor held on a staff approximately 2.5 m above the ground. The total magnetic field, location, and time of acquisition were recorded for every station. Because the

Earth's magnetic field is subject to several time variations, of which the daily or diurnal variation is principle, a profile is usually preceded with a reading (and time) at an arbitrary local base station. The base station is then read again after completion of the profile. Since measurements along each of the profiles were acquired in less than an hour, the drift could be approximated as a simple linear function and readily corrected. Of the 19 VLF profiles conducted, 18 had magnetic data collected simultaneously.

Vertical Electrical Soundings (VES)

A Vertical Electrical Sounding (VES) is a type of resistivity measurement that measures electrical resistivity as a function of depth at a fixed point, assuming sub-horizontal layering (Telford et al., 1990). Changes in resistivity were used to identify changes in lithology and the location of the water table below the measurement point. Individual VES measurements were integrated to produce a geoelectric cross section from which the lithologic and hydrologic variations across a profile path were constructed.

The VES measurements taken in the field (using IRIS SYSCAL R2) were performed using a four-electrode expanding Schlumberger Array. In this array, the two potential electrodes (M and N) are kept fixed for three to five different expansions of the current electrodes (A and B). Because the current electrode spacing (A–B) is the primary control of the depth distribution of current lines (in addition to the resistivity layering structure), the AB or AB/2 distance is the reference for each reading. We used AB/2 spacings that increased from 1.0 m to 147 m in equally spaced intervals on a logarithmic scale, six spacings per decade. MN spacings were always 1/5 or less of the AB spacings. At the point where the measuring electrodes were expanded, another reading was taken without moving the current electrodes, providing data overlap. The first stage of treatment of the Schlumberger VES data was to make a smooth curve (on a log resistivity versus log of AB/2) either by adjusting segments up or down or by averaging the overlap points if the offset is small (<1%). Finally, an inversion routine was used to convert field resistivity versus AB/2, to interpreted resistivity versus depth. A total of 10 soundings were taken near a central location of the HF (Fig. 2B).

Isotopic methods

Twenty-two groundwater samples were collected (winter 2010) from shallow (depth: 40–70 m) production wells penetrating the Regional Aquifer and six samples were collected (spring 2011) from the springs in the foothills of the San Bernardino Mountains. The hydrogen and oxygen isotopic ratios were analyzed using a Picarro Cavity Ring-down Spectroscopy (CRDS) laser system (Lehmann et al., 2009) at ISOTECH Laboratories in Champaign, Illinois. Hydrogen and oxygen stable isotope ratios are reported using conventional delta (δ) notation, in units of per mil (‰) deviation relative to the Vienna Standard Mean Ocean Water (V-SMOW), whereby

$$\delta(‰)=[(R_{sample}/R_{standard})-1]\times 10^3 \quad (1)$$

and $R = {}^2H/{}^1H$, or ${}^{18}O/{}^{16}$ (Coplen, 1996). Reproducibility of δ values for δ D is ±1‰ and that of δ ^{18}O is ±0.2‰. The isotopic analyses can be used to investigate the source(s) of groundwater and to identify mixing trends between these sources (e.g.,Clark and Fritz, 1997). In general, contrasts in these isotopic compositions across structural features such as uplifts and faults could be indicative of the role these features play as barriers to the lateral groundwater flow that tends to homogenize the groundwater composition on both sides of the fault.

Geographic Information System (GIS) analyses

Analysis of the collected data for this project involved the generation of a database for data integration, analysis, and visualization; this was made possible by using Arc Info GIS and Arc Spatial Database Engine (SDE). The generated database incorporates all relevant co-registered digital mosaics with a unified projection (World Geodetic System [WGS] 84) covering the southern Mojave Desert: (1) geologic map (scale 1:250,000); (2) LiDAR (swath width: 1.5 km; spatial resolution: 1 m) spanning the approximate length of the HF; (3) well data, provided by Mojave Water Agency, including one or more of the following parameters: well location, well name, well type (i.e., production, commercial, domestic, municipal), depth to water table, well elevation, and hydrogen and oxygen stable isotopic compositions; (4) IKONOS and eoEye-1 satellite scenes (spatial resolution 1–5 m); (5) California Alquist-Priolo (AP) Earthquake Fault Zone maps (scale 1:24,000) (Bryant and Hart, 2007); (6) VES soundings in the immediate vicinity of the HF; and (7) VLF and magnetic profiles acquired in tandem for dextral faults and intersecting fault splays in and near the HF.

DATA AND FINDINGS

VLF and magnetic data

VLF data can be examined directly on tilt-angle versus profile distance plots. Conductors are located where the tilt angle changes sign (zero crossover). Alternatively, simple filtering can be applied to remove short-wavelength features (e.g., single-station anomalies) and to shift the peaks by 90°. The four-point linear Fraser filter (Fraser, 1969) is widely used to accomplish this; it shifts the curve so that positive peaks are directly over the conductors. For a given sequence ($X1$, …, Xn) of tilt angle measurements taken along the same profile at a regular station interval, the Fraser filtered value ($F1$) was estimated as: $F1 = (X3 + X4) - (X1 + X2)$ and plotted half way between positions of $X2$ and$X3$ measurements. Fraser (1969) pointed out that in optimal conditions a filtered peak of 5% may be significant, and that values can go as high as 100%. For this survey, we conservatively regarded Fraser-filtered peaks above 20% as significant, even though on some days the threshold could have been 10% or 15%, as verified by comparing profiles repeated at the same location on different dates.

Seven representative profiles are detailed in the discussion; five have magnetic data. Three of the profiles were acquired along the northern portion of the HF, and the remaining four were acquired along the southern portion of the fault. Fig. 3A–F shows the VLF and magnetic profiles at three locations (Fig. 2B: P1104, P1103, P0209) along the northern portion of the HF. A LiDAR swath or high-resolution GeoEye-1 imagery was used as a background image and faults were mapped locally (dashed lines) from these images. Several scenes also have the California AP as an added background; these zones outline buffer areas where movement along active faults has been measured since 1971 by the California Geological Survey (see www.consrv.ca.gov). The VLF responses shown in Fig. 3A–C, together with the magnetic measurements that were acquired simultaneously along the same profiles, were graphed in Fig. 3D–F. Similar displays (Fig. 4A–C) and response graphs (Fig. 4D–G) were generated for the VLF and magnetic profiles (Fig. 2B: P3001, P1003, P0105, and P1105) located along the southern portion of the HF in the more mountainous region southwest of Lucerne Valley.

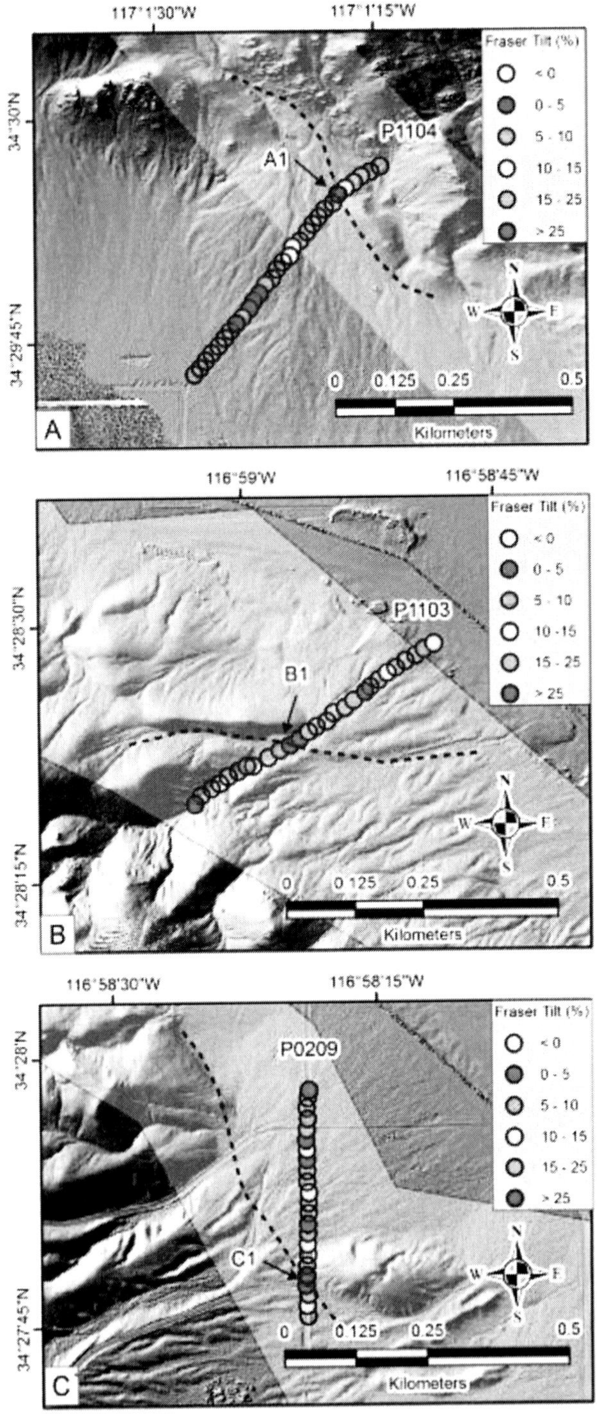
117°1'30"W
117°1'15"W
34°30'N
34°29'45"N
Fraser Tilt (%)
< 0
0 - 5
5 - 10
10 - 15
15 - 25
> 25
P1104
A1
N
W
E
S
0
0.125
0.25
0.5
Kilometers
A
116°59'W
116°58'45"W
34°28'30"N
34°28'15"N
P1103
B1
B
116°58'30"W
116°58'15"W
34°28'N
34°27'45"N
P0209
C1
C

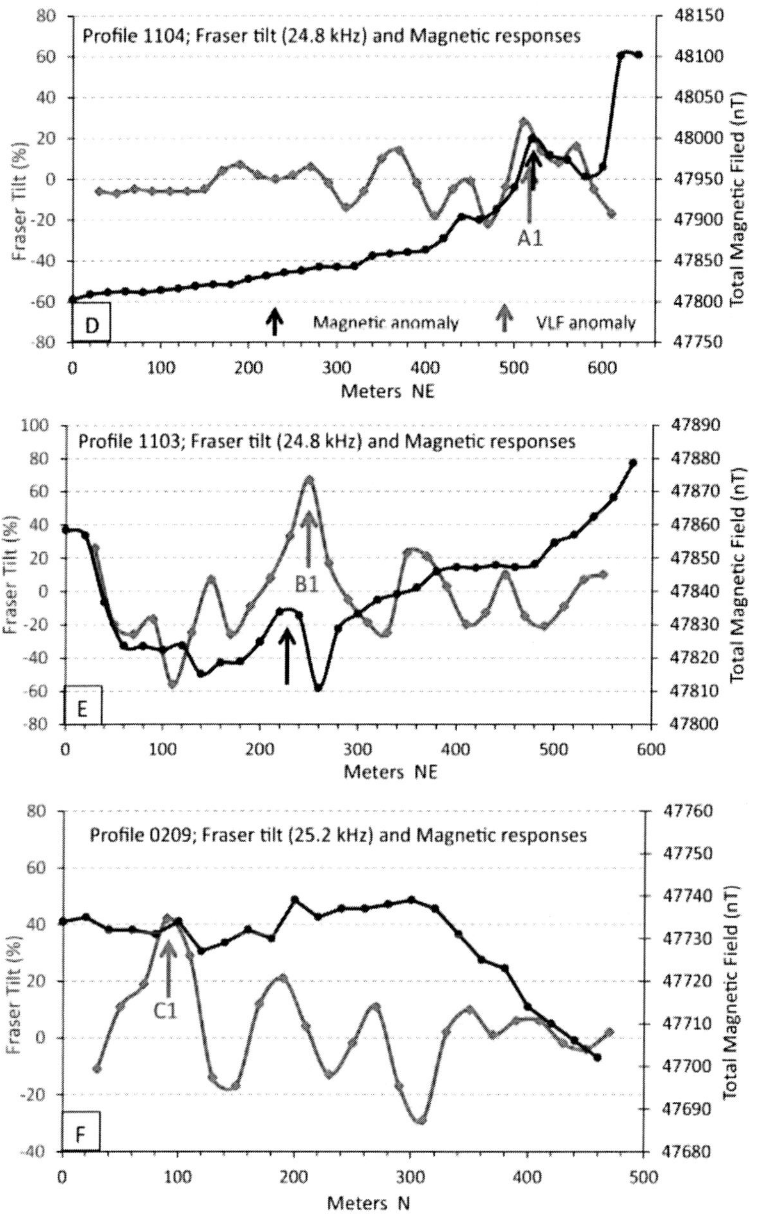

Figure 3. Selected VLF and magnetic profiles and corresponding data plots for the northwest portion of the HF. (A–C) VLF responses (Fraser tilt ≥0%: colored circles; Fraser tilt < 0%: hollow circles) over the fault locations (dashed black line) within California Alquist-Priolo (AP) zones (light blue area). (D–F) The data plots (red lines: Fraser tilt [%]; black lines: magnetic responses [nT]) for each of the measured profiles. Appreciable VLF and magnetic responses (marked by arrows) correspond to known (black dashed lines) fault locations in (A–C).

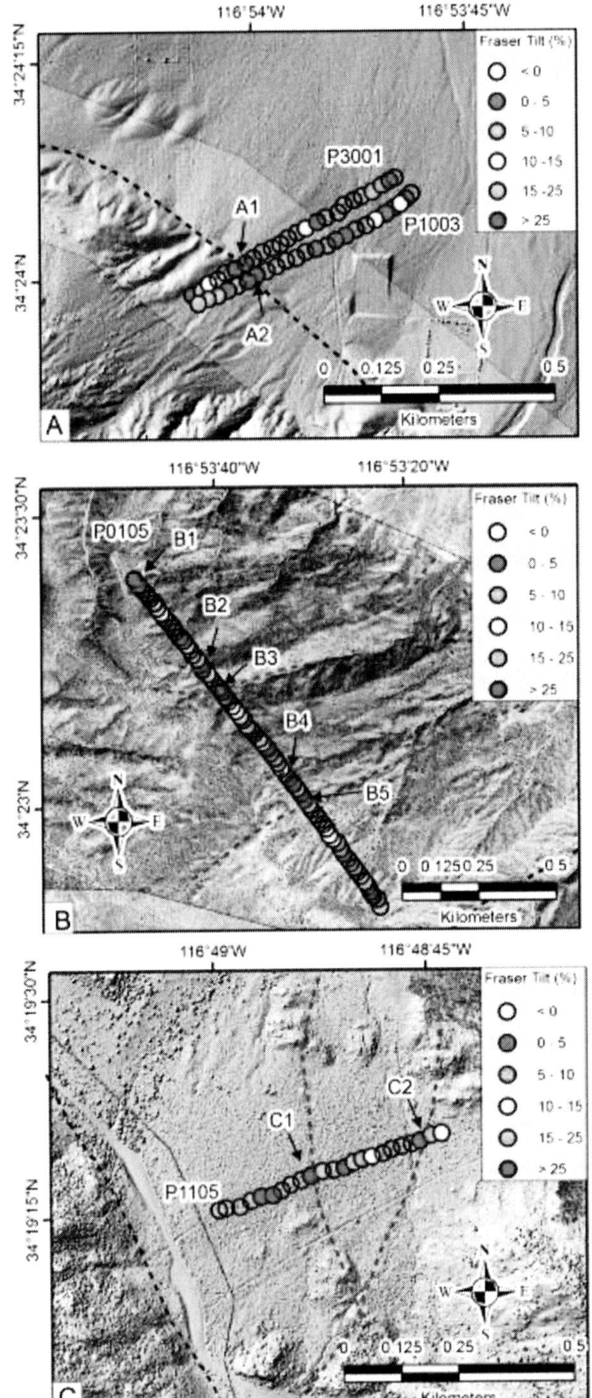
116°54'W
116°53'45"W
34°24'15"N
34°24'N
Fraser Tilt (%)
< 0
0 - 5
5 - 10
10 - 15
15 - 25
> 25
P3001
A1
P1003
A2
0
0.125
0.25
0.5
Kilometers
A
116°53'40"W
116°53'20"W
34°23'30"N
34°23'N
P0105
B1
B2
B3
B4
B5
B
116°49'W
116°48'45"W
34°19'30"N
34°19'15"N
C1
C2
P1105
C
N
S
E
W

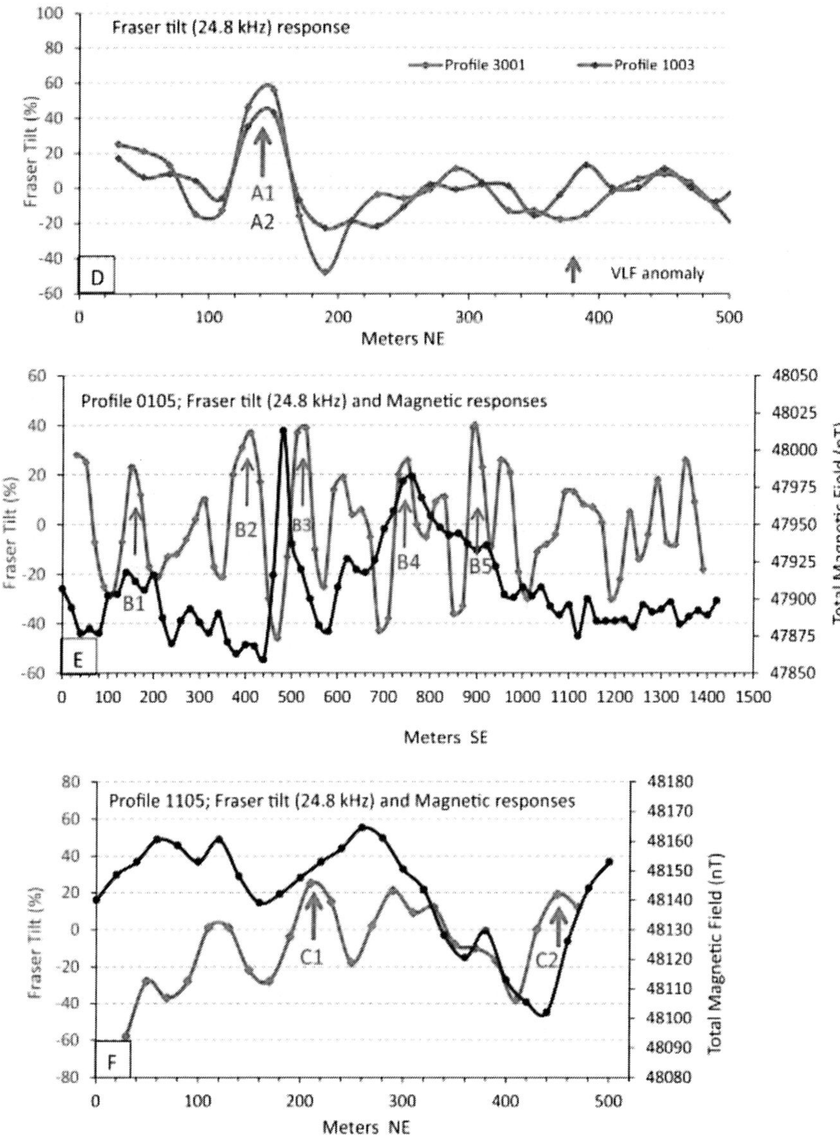

Figure 4. Selected VLF and magnetic profiles and corresponding data plots for the southeast portion of the HF (black dashed line), fault splays (red dashed line) mapped from LiDAR and Geoeye-1 imagery, and California Alquist-Priolo (AP) zones (light blue area). (D–F) The data plots (red lines: Fraser tilt [%]; black lines: magnetic response [nT]) for each of the measured profiles. Appreciable responses (marked by arrows) correspond to known fault locations (black dashed lines) or LiDAR- and satellite-based inferred fault locations (red dashed lines) on (A–C).

Profile P1104 was performed in a box canyon, moving west to east (Fig. 3A and D). The modest VLF peak (Fraser Tilt [FT] 30%) aligns well with the main trace of the HF, which is well defined in the LiDAR dataset, and lies well within the mapped AP zone (light blue shaded areas outlined with thin black dotted lines). The magnetic response shows a change from a smooth, gentle gradient in the southwest to steeper gradients and a local maximum at the location of the FT VLF maximum. This is an indication of the juxtaposition of two different basement lithologies and/or shallow basement to the northeast and deeper basement to the southwest. Profile P1103 (Fig. 3B and E) shows good responses with both magnetic and VLF methods. The strong VLF anomaly (FT 65%) aligns with a clear dipolar magnetic response. Both responses align well with a mapped fault that is shown on the LiDAR image and is also identified on geologic maps. This fault also lies well within the mapped AP zone for the HF. This fault is most likely a branch of the primary HF that "kinks" through the large ridge, with the main trace obscured to the east, or possibly still buried beneath the relatively shallow valley sediments. Another strong anomaly at the beginning of the profile (southwest end) aligns with a major fault, which is most likely the reason for the large depression seen in the LiDAR image. Profile P0209 (Fig. 3C and F) shows a strong VLF response (FT 45%) that aligns well with a mapped portion (AP zoning and local geologic maps) of the HF, which curves through the canyon in this area. The magnetic profile does not show any significant trend changes or anomalies at this position.

Fig. 4A–C shows four VLF profiles, two of which also have magnetic data. Profiles 3001 and 1003 (P3001, P1003; Fig. 4A and D) are repeat profiles, done each year in approximately the same location. P1003 was performed in summer of 2009, while P3001 was performed earlier in the summer of 2011 (after a wetter spring). P3001 shows a single, well-defined peak with a FT of 56% very near the fault plane, while P1003 shows a similar peak with a FT of 43%. Both peaks also align very well with the mapped AP zone for this area of the HF. The profile from 2011 showed a significantly higher response than that from 2009 (FT 13% higher). This is most likely due to the timing of the profile and the increased amount of groundwater still present in the fault plane when measured. Both profiles show high FT values at the start (southwest end), indicating another possible fault splay to the left and higher up the slope of this ridge.

Profile 0105 (P0105; Fig. 4B and E) runs parallel to the HF and crosses at high angles a series of suspected fault splays that intersect the HF itself. In each case, an appreciable peak (FT $\geq$ 20%) is seen as the profile crosses the fault traces, indicating the presence of shallow sub-vertical conductive sheets, most likely groundwater within high-angle fault splays. If these splays connect to the HF, they could possibly collect additional groundwater to add to that already channeled by the main fault. This magnetic profile shows much more variability than most of the other measured profiles. Bedrock is very shallow, and there are obvious lithologic variations and topographic effects along the profile that probably cause most of the observed abrupt magnetic changes. P1105 crosses two suspected splays (Fig. 4C: red dashed lines) of the HF (mapped as a black dashed line) in the far south. The fault traces here are fairly well defined, and two appreciable VLF peaks are observed as each fault trace is crossed, again indicating the presence of a shallow, conductive material. The magnetic profile for this transect is not diagnostic of faulting (Fig. 4F). Our VLF data are consistent with the interpretation that the HF is providing enhanced groundwater flow opportunities along strike.

VES data

The locations as well as the inversion results (number of layers, resistivity, thickness, depth and rms fit values) of the 10 VESs are shown in Fig. 5A. VESs 9, 4, 6, and 7 were aligned parallel and offset to the east of the fault; VESs 10, 1, 5, and 8 were located parallel to the west side of the fault. In addition, soundings 2 and 3 were positioned west of VES 1, thus creating a west-east geoelectric section (VESs 3, 2, 1, and 4) across the fault.

East of the fault, the four soundings show a strong similarity in curve types, each with two maxima. The northernmost sounding (VES 9) shows the first maximum is at the surface and becomes progressively deeper to the south. The intervening minimum and the second maximum also become progressively deeper; overall resistivities increase as the soundings progress south-southeast and up the alluvial fan to higher elevations. Each of these four soundings ends with a steeply declining branch, indicating that the conductive, saturated zone is being detected. This conductive zone is observed at progressively greater depths to the south-southeast. The two maxima are likely coarse-grained vadose zone sediments sandwiching a more conductive fine-grained unit.

A very different trend is observed west of the fault. In general, a much lower trend in resistivities is observed in each of the soundings when compared to similarly placed soundings east of the fault. They are similar only in that the resistivities rise, moving north to south up the alluvial fan surface. The right (deepest) branch of the log-log VES curves 1, 5, and 8 shows an increase in resistivity. This is in contrast to the case on the east

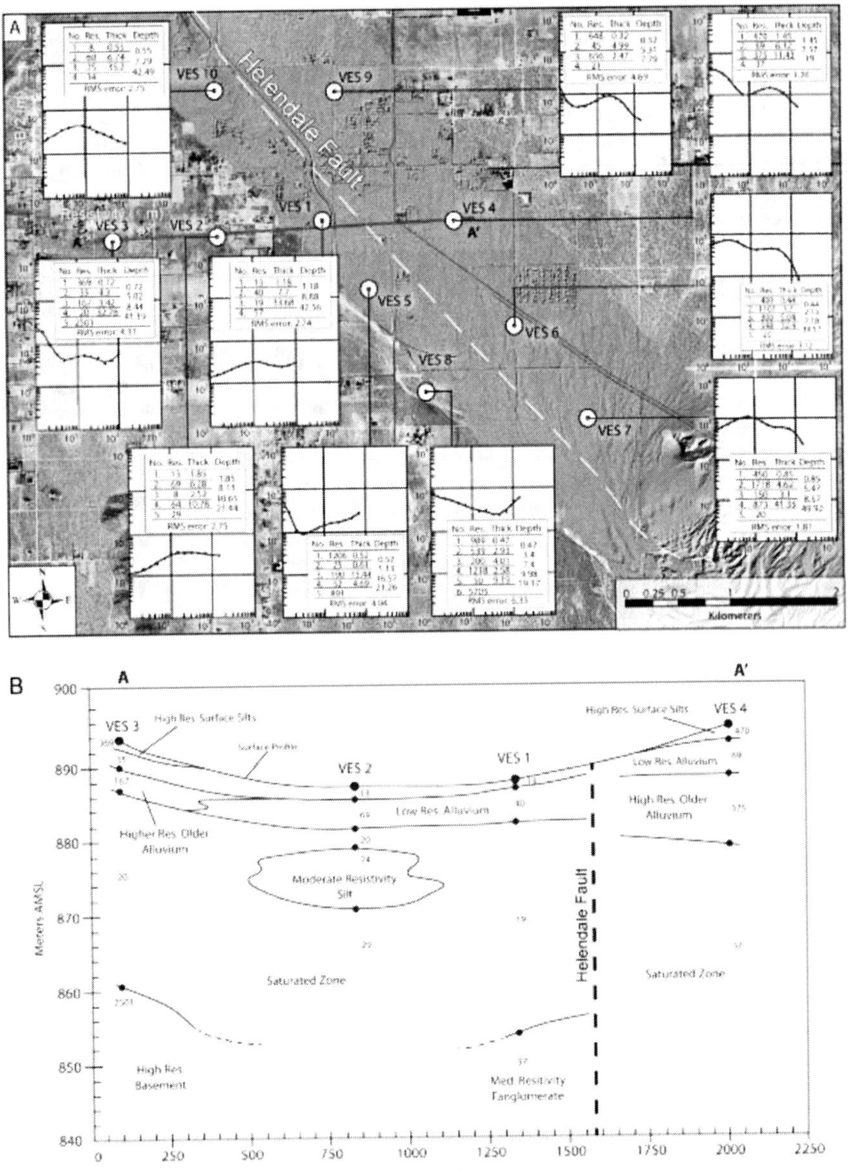

Figure 5. (A) VES location, apparent resistivity, thickness, and distribution of saturated and unsaturated rock units in areas west (VESs 10, 3, 2, 1, 5, and 8) and east (VESs 9, 4, 6, and 7) of the HF (white dashed line). Also shown is the distribution of LiDAR (wavelength: 1047×10^{-9} m) data. (B) Geoelectric West-East trending cross-section A-A′ showing the VES (1–4) locations and apparent resistivity (Ω m), thickness (m a.m.s.l), and distribution of the saturated and unsaturated rock units. The top of the saturated zone represents the depth to water table. The vertical dashed line represents the inferred location of the HF.

side of the fault; the most plausible interpretation is that bedrock was encountered at fairly shallow depths. These sections were conducted along the strike of the topographic ridge to the southeast, and would appear to make a strong case for the continuation of that bedrock ridge in the subsurface all the way north to Lucerne Valley. If this is the case, then this bedrock ridge just to the west of the HF is likely what is blocking the eastward flow of groundwater across the fault. Flow is then only possible through a passage that is on the order of 1.5 km wide at the valley bottom, starting north of VES 1 and terminating against a ridge slightly north of VES 10 (this is the same ridge seen in Fig. 2A).

Soundings 3, 2, 1, and 4 (west-east row), show a clear change across the fault between soundings 1 and 4. They also show the low resistivity of the two central soundings (VES 2 and VES 1) where the surface sediments consisted of fine silts and clays, typical of the valley floor. The surface material at VES 3 was coarser, which typically causes higher resistivity, but the deeper resistivities were also low, as with soundings 1 and 2. The deepest segment of the curve for sounding 3 is a rising branch, possibly indicating another area of shallow bedrock. Another, much smaller, ridge is seen to the southeast of sounding 3 and may continue to the northwest beneath VES 3. It is reasonable to assume that this ridge, like the ridge that parallels the HF, continues northwest in the shallow subsurface before outcropping again to the northwest.

The geoelectric cross-section (Fig. 5B) along line A–A′ (Fig. 5A) shows that the vertical resistivity structure at VES 4 is different from that of VESs 1–3. West of the fault, the water table is interpreted at about 7–10 m depth, and a short distance east of the fault it is double that value. East of the fault, a thick, high-resistivity (375 Ω m) unsaturated zone is present, whereas a much lower resistivity, probably reflecting finer-grained sediments, is observed west of the fault. Two of the VESs to the west appear to have reached high-resistivity basement rock at depths of about 40 m. This supports the hypothesis of a shallow bedrock sill joining the topographic ridges to the northwest and southeast.

Analysis of our resistivity data indicate the presence of high resistivities west of the fault at the right end of some of these VES curves, characteristic of shallow bedrock; east of the HF, the VES curves end at low values typical of aquifer units, not bedrock. These high resistivities occur along the strike of a prominent bedrock ridge, which runs parallel to the HF and dips below the surface along the valley floor. Further west of the fault, shallow resistivity is much lower and fits with mapped geologic units of young alluvium overlying older, higher-resistivity alluvium. Farther west still, another deep resistivity high is seen, again along strike of a bedrock ridge to the southeast. It is quite possible that these ridges are generally preventing groundwater from flowing east through the faulted zone, except across open “windows” or saddles between the shallow bedrock ridges that trend parallel to the fault.

Stable isotope data

Isotopic analyses (δD and $\delta^{18}O$) were conducted on groundwater samples collected from productive wells and springs. The distribution of the well and spring samples is given in Fig. 6, and the isotopic analyses of these samples are presented in Table 1 andFig. 7; well and spring samples are labeled “mwa” and “wmu” samples, respectively. Inspection of Table 1 and Fig. 6 and Fig. 7 show that many of the samples plot to the right of the global meteoric, which suggests that they were subjected to variable degrees of evaporation. Previous studies in the LVGS and surrounding area suggested that the aquifers in the investigated areas have been recharged in part in previous wet and cool climatic periods (Smith et al., 1992 and Izbicki, 2004). The analyzed groundwater samples were classified into three groups based on their sampling location with respect to the HF: (1) Group I (mwa: 2, 14–16; wmu: 1, 2) were collected west of the fault from the FVGS; (2) Group II (mwa: 3, 4, 6–10, 17, 18, 20–24; wmu: 4) were sampled from the LVGS; and Group III (mwa: 1, 12, 13, 19; wmu: 3, 5, 6) were collected from locations proximal (<1.5 km) to the HF.

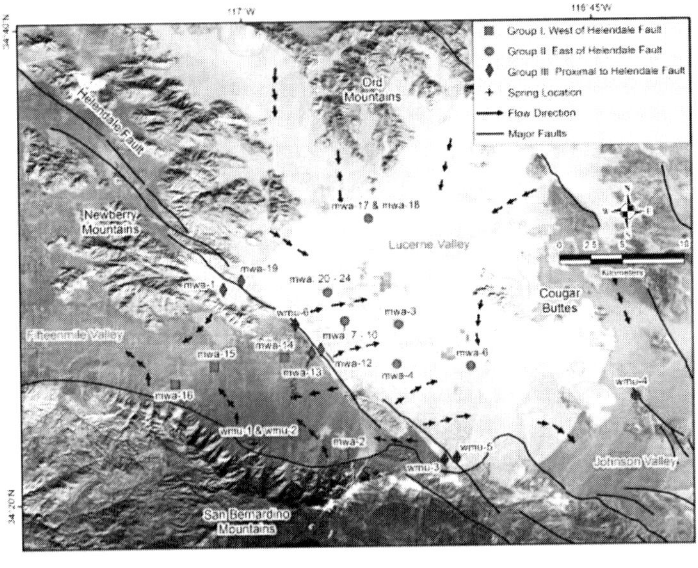

Figure 6. Locations of wells (name designation: mwa) and springs (black cross; name designation: wmu) sampled for isotopic analyses (O, H) grouped into three main groups (I: West of HF [blue squares]; II: East of HF [red circles]; III: proximal to HF [green diamonds]). Also shown are (1) Groundwater sub-basins (i.e., Lucerne Valley, Fifteenmile Valley, and Johnson Valley); and (2) groundwater flow direction within the sub-basins (Stamos and Predmore, 1995, ESA, 2004 and MWA, 2005).

Table 1. Isotopic data for groundwater samples from wells and springs from the Fifteenmile Valley and Johnson Valley groundwater sub-basins.

Name	Groundwater sub-basin	Latitude (DD[a])	Longitude (DD)	Elevation (m)	Well description	Sample depth (m)	DTW[b](m)	δD (‰)	$\delta^{18}O$ (‰)
wmu-1	Fifteenmile Valley	34.3758	−116.9929	1356.1	Mountain Front Spring	N/A	0	−85.6	−11.8
wmu-2	Fifteenmile Valley	34.3747	−116.9945	1356.1	Mountain Front Spring	N/A	0	−85.2	−11.4
wmu-3	Lucerne Valley	34.3628	−116.8575	1398.7	Mountain Front Spring	N/A	0	−88.3	−11.7
wmu-4	Johnson Valley	34.4058	−116.7208	966.8	Valley Spring	N/A	0	−96.6	−12.5
wmu-5	Lucerne Valley	34.3641	−116.7488	1270.1	Mountain Front Spring	N/A	0	−86.5	−10.4
wmu-6	Fifteenmile Valley	34.4582	−116.9636	875.1	Valley Spring	N/A	0	−84.7	−10.5
mwa-1	Lucerne Valley	34.4830	−117.0150	916.5	Domestic Well	–	–	−85.3	−11.0
mwa-2	Fifteenmile Valley	34.3669	−116.9295	1263.4	Production Well	–	–	−87.5	−11.6
mwa-3	Lucerne Valley	34.4575	−116.8893	891.8	Production Well	–	–	−91.6	−12.2
mwa-4	Lucerne Valley	34.4298	−116.8910	925.7	Domestic Well	69	50.3	−93.2	−12.2
mwa-6	Lucerne Valley	34.4280	−116.8384	933.6	Domestic Well	–	–	−91.8	−11.9
mwa-7	Lucerne Valley	34.4605	−116.9283	876.9	Municipal Well	55	47.2	−97.5	−11.6
mwa-8	Lucerne Valley	34.4605	−116.9283	876.9	Municipal Well	55	46.3	−100.7	−12.8
mwa-9	Lucerne Valley	34.4605	−116.9283	876.9	Municipal Well	44	41.5	−92.0	−12.1

mwa-10	Lucerne Valley	34.4605	−116.9283	876.9	Municipal Well	55	46.6	−91.8	−12.5
mwa-12	Lucerne Valley	34.4402	−116.9456	904.3	Commercial Well	–	–	−91.2	−12.3
mwa-13	Lucerne Valley	34.4380	−116.9520	916.8	Domestic Well	–	–	−91.4	−12.4
mwa-14	Fifteenmile Valley	34.4352	−116.9714	909.8	Domestic Well	–	–	−88.2	−12.2
mwa-15	Fifteenmile Valley	34.4290	−117.0219	1033.3	Municipal Well	50.	25.3	−87.9	−11.9
mwa-16	Fifteenmile Valley	34.4170	−117.0500	877.8	Domestic Well	–	–	−86.4	−11.7
mwa-17	Lucerne Valley	34.5318	−116.9104	877.8	Municipal Well	55	46.3	−95.6	−10.5
mwa-18	Lucerne Valley	34.5318	−116.9104	884.5	Municipal Well	55	46.3	−97.7	−11.0
mwa-19	Lucerne Valley	34.4890	−117.0020	884.5	Domestic Well	–	–	−94.1	−12.6
mwa-20	Lucerne Valley	34.4805	−116.9405	869.9	Municipal Well	44	39.0	−96.7	−11.8
mwa-21	Lucerne Valley	34.4805	−116.9405	869.9	Municipal Well	44	39.3	−98.3	−12.4
mwa-22	Lucerne Valley	34.4805	−116.9405	869.9	Municipal Well	53	39.0	−93.9	−12.4
mwa-23	Lucerne Valley	34.4805	−116.9405	869.9	Municipal Well	53	39.3	−93.3	−12.8
mwa-24	Lucerne Valley	34.4805	−116.9405	869.9	Municipal Well	38	32.0	−94.3	−12.8

[a]DD – decimal degrees.
[b]DTW – depth to water table.

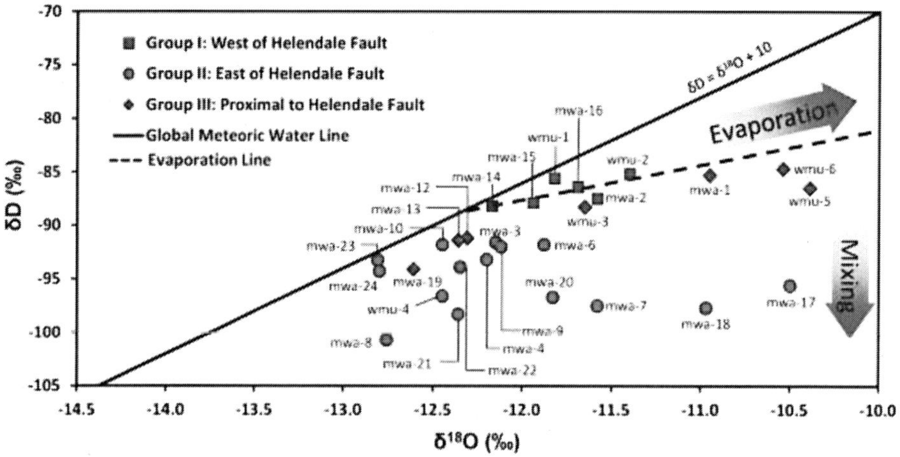

Figure 7. Comparison between stable isotope ratios (hydrogen [δD] versus oxygen [δ^{18}O]) for groundwater samples from wells and springs west, east, and proximal to the HF. Also shown are the evaporation line for Group I samples, and its intersection (δD: −89‰; δ^{18}O: −12.3‰) with the Global Meteoric Line (δD = δ^{18}O + 10;Craig, 1961).

Group I samples (δD: −85.2‰ to −88.2‰, average: −86.8‰; δ^{18}O: −11.4‰ to −12.2‰, average: −11.8 ‰) plot along an evaporation line (Fig. 7; EVL) with a source whose composition (δD: −89‰; δ^{18}O −12.3‰) is given by the intersection of the evaporation and global meteoric lines. The inferred composition for the source is similar to that for groundwater (δD: −84 ‰) from wells recharged by runoff from the higher altitudes of the San Bernardino and San Gabriel Mountains (Izbicki, 2004). Because two of these samples (wmu: 1 and 2) were collected from springs in the foothills of the San Bernardino Mountains, we suggest that the source of Group I samples is largely snowmelt from the mountains. This suggestion is consistent with earlier findings (USGS, 2004) indicating that the Regional Aquifer is being recharged by runoff and groundwater flow from the San Bernardino Mountains in the south. Irrespective of the source, the data clearly indicates that the water has undergone significant evaporation, which can take place during precipitation, in the unsaturated zone, or due to excessive use and “return flow” (Murad and Krishnamurthy, 2008).

Group II samples show a wide range of variation in their isotopic composition and are more depleted (δD: −91.6‰ to −100.7‰, average: −95.0‰; δ^{18}O: −10.5‰ to −12.8‰, average: −12.1‰) compared to Group I samples. It is likely that the LVGS is recharged largely from precipitation over the northern mountains (Ord, Rodman, and Stoddard mountains)

(Blazevic et al., 2008) and that the groundwater flow is toward the southeast (Fig. 1 and Fig. 6). It is commonly thought that the HF acts as a barrier to lateral groundwater flow, which would prevent groundwater from the FVGS from mixing with that of the LVGS. This does not appear to be the case. Group II samples plot toward the Group I evaporation line (Fig. 7). These wells typically are east of the HF and fall along the east-southeast groundwater flow path in the LVGS. This indicates that groundwater mixing is occurring within the LVGS. The isotopic composition of Group III samples is consistent with an explanation that calls for various degrees of evaporation and mixing for Group I and II compositions. Samples collected from wells distant from the HF and proximal to their mountainous source areas (mwa: 17 and 18; Fig. 6 and Fig. 7) are the most separated from the evaporation line and thus could be the Group II samples least affected by mixing.

Carbon-14 ages (up to 43,000 year BP) reported for groundwater from the FVGS and the LVGS indicate that these aquifers were recharged at least in part in previous wet climatic periods (Izbicki and Michel, 2004). The depleted nature of Group II samples could be related to the fact that it is dominated by fossil groundwater; this suggestion is supported by the fact that the precipitation over the FVGS source areas (San Bernardino Mountains) far exceeds that over the LVGS source areas (Newberry and Ord mountains) (Izbicki, 2004 and USGS, 2004). This could potentially lead to excessive flushing of the fossil water in the FVGS, and less flushing in the case of the LVGS.

Group III samples were taken from areas on either side of, and proximal to, the HF. In each case, the samples plot along, or close to but below, the evaporation line for Group I samples; their composition is slightly more depleted than Group I samples, yet more enriched than Group II samples (δD: −84.7‰ to −94.1‰, average: −88.8‰; $\delta^{18}O$: −10.4‰ to −12.6‰; average: −11.5‰). This indicates that their composition is largely the same as that of Group I samples, the San Bernardino Mountain recharge waters; however, they show evidence for mixing with the Group II waters, the LVGS groundwater. This is to be expected if the HF is channeling Group I groundwater from the San Bernardino Mountains; limited mixing with Group II will occur proximal to the fault, and more extensive mixing will take place along the groundwater flow direction(s) within the LVGS.

Introduction of Group I groundwater in the LVGS is not only occurring by channeling groundwater along the HF, but possibly by groundwater flow across open "windows" or saddles between the shallow bedrock ridges that trend parallel to the fault. Two of these windows occur in the vicinity of wells mwa-13 and mwa-12, as well as at springs wmu-3 and wmu-5. In both cases, the bedrock ridge that cuts across the majority of the valley floor is unexposed at the surface. Since water levels west of the fault are relatively shallow, it is plausible that water is able to spill over the bedrock barrier in these locations, facilitating mixing between the aquifers. Regardless of which hypothesis is adopted, groundwater samples from the LVGS that have mixed isotopic compositions must have been derived, at least in part, from recharge waters in areas west of the fault because if an impermeable barrier were preventing lateral flow, these anomalous mixed composition values could not be explained.

If the HF is channeling Group I groundwater from the San Bernardino Mountains, one would expect that groundwater from areas proximal to the HF to have higher carbon-14 activities compared to those collected from areas distant from the fault. Inspection of reported carbon-14 activities expressed in percent modern carbon (pmc) (Izbicki and Michel, 2004) shows that is indeed the case; pmc values are high in samples collected from areas proximal to the HF (distance: 2–3 km; pmc: 69) and progressively decrease in values with increasing distance from the HF (distance: ~15 km; pmc: 36) along postulated groundwater flow lines (Fig. 8). Similarly, high pmc values (pmc: 41–86) were reported from areas proximal and downgradient from the Camp Rock Fault (Fig. 8). One would also expect to detect tritium downgradient from the HF if it is channeling Group I groundwater from the San Bernardino Mountains. Tritium was detected in observation wells tapping the Regional Aquifer and underlying deposits downgradient from the HF; it was also detected downgradient from the Camp Rock Fault (Izbicki and Michel, 2004).

RESULTS AND REGIONAL IMPLICATIONS

Our data (VLF, VES, O, H isotopic composition) and reported pmc values (Izbicki and Michel, 2004) are consistent with the interpretation that the HF is providing enhanced groundwater flow opportunities along strike, and channeling groundwater from the higher-elevation areas in the south to the lowland areas in the central and northern portions of the MRGB and MGB. The VLF data shows that the majority of locations along the HF are conductive, indicating that the fault plane contains groundwater even at higher elevations on the southern ridges.

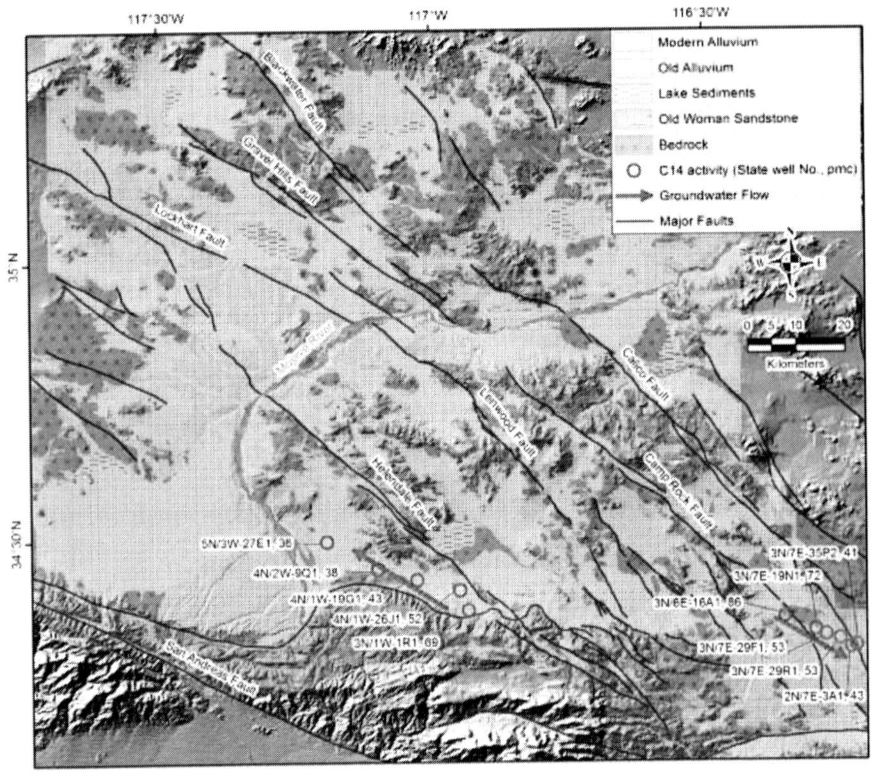

Figure 8. Simplified geologic map (modified from MWA, 2007) with hillshade applied to accentuate topographic features showing the distribution of uplifted basement outcrops along and proximal to the major dextral faults in the Mojave Desert block. Also shown are sample locations, pmc values, and groundwater flow directions from Izbicki and Michel (2004).

Our groundwater samples show evidence for varying degrees of evaporation and mixing. Samples taken in areas to the east of the HF (Group II), especially those from wells along the east-southeast groundwater flow path in the LVGS groundwater and those that were sampled proximal to the HF (Group III), show limited mixing and more extensive mixing along groundwater flow path within the LVGS. One way for such mixing to occur is for the fault to channel groundwater through its fracture planes, many of which are observed within the bedrock ridges. If the HF acted as an impermeable barrier preventing lateral flow, these mixed-composition values from the LVGS could not be explained, since these samples must have been derived, at least in part, from recharge waters in areas west of the fault.

If the HF is not a barrier for lateral groundwater flow, then what is causing the observed change in water level across the fault? West of the fault, groundwater is found an average of 5–10 m below surface level, while to the east of the fault, groundwater depth averages 45 m below surface level (Smith and Pimentel, 1998 and Stamos et al., 2003). Our geophysical data (VES) suggest that the groundwater drop across the HF can be at least partially attributed to the shallow bedrock ridges that restrict or impede the groundwater from flowing east across the fault. The presence of basement uplifts running parallel and proximal to the HF could be also inferred from the trail of uplifted basement outcrops that are parallel to, and bound by, the fault traces (Fig. 1 and Fig. 8).

Our findings for the HF might be relevant to many of the other major dextral faults within the Mojave Desert block. As is the case with the HF, the Lenwood, Camp Rock, Calico, and Granite mountains and Johnson Valley faults are decorated by uplifted basement outcrops along the majority of their lengths (Fig. 8). A strong relationship between topography and the strain domains has been observed within the Mojave Desert block. Many of the elevated areas, previously referred to as uplifts, in the Mojave Desert block are experiencing oblique shear resulting from dextral strike-slip shear and shortening (Dokka and Macaluso, 2001).

This study highlights the potential for identifying similar fault zones throughout the Mojave Desert region, since it is likely that these faults, especially those with exposed bedrock ridges trending along their length, will exhibit similar behavior; they channel groundwater along their length from source areas in the highlands to populated areas in the lowlands. From a groundwater supply standpoint, this is important because it indicates additional resources may be available for the dry, low-precipitation areas, which are increasing in population each year. Caution should be exercised in generalizing our findings. Studies similar to the ones advanced here should investigate whether the faults that experience a large throw and generate a wide gouge zone, like the San Andreas Fault, are channeling groundwater along their strike like the HF and in the same time acting as barriers for lateral groundwater flow as originally perceived.

ACKNOWLEDGEMENTS

We thank the Mojave Water Agency for the logistic support provided. We thank Phil Armstrong and the California State University, Fullerton, Department of Geology for their support in providing field equipment and technical support. We thank R.V. Krishnamurthy for fruitful discussions regarding the isotopic data results. Funding for this project was supplied by the Mojave Water Agency and California State University, Fullerton (grant no. S4879-WMU).

REFERENCES

1. Barton, C.A., Zoback, M.D., Moos, D., 1995. Fluid-flow along potentially active faults in crystalline rock. Geology 23, 683–686.
2. Blazevic, M., Laton, W.R., Foster, J., 2008. Geologic Insight to Lucerne Valley Groundwater Basin. California State University, Fullerton.
3. Bense, V.F., Person, M.A., 2006. Faults as conduit-barrier systems to fluid flow in siliciclastic sedimentary aquifers. Water Resour. Res. 42, W05421, http://dx.doi.org/10.1029/2005WR004480.
4. Bense, V.F., Van Balen, R.T., DeVries, J.J., 2003. The impact of faults on the hydrogeological conditions in the Roer Valley Rift System: an overview. Neth. J. Geosci. 82 (1), 41–54.
5. Bryant, W., Hart, E., 2007. Fault-rupture Hazard Zones in California: Alquist-Priolo Earthquake Fault Zoning Act with Index to Earthquake Fault Zones Maps. Special publication no. 42, California Geological Survey., pp. 1–42.
6. Caine, J.S., Evans, J.P., Forster, C.B., 1996. Fault zone architecture and permeability structure. Geology 24, 1025–1028.
7. Caine, J.S., Minor, S.A., 2009. Structural and geochemical characteristics of faulted sediments and inferences on the role of water in deformation, Rio Grande Rift New Mexico. Geol. Soc. Am. Bull. 121 (9–10), 1325–1340.
8. Celico, F., Petrella, E., Celico, P., 2006. Hydrogeological behaviour of some fault zones in a carbonate aquifer of Southern Italy: an experimentally based model. Terra Nova 18, 308–313.
9. Clark, I.D., Fritz, P., 1997. Environmental Isotopes In Hydrogeology. CRC Press/Lewis Publishers, Boca Raton, FL. Coplen, T.P., 1996. New guidelines for reporting stable hydrogen, carbon and oxygen isotope-ratio data. Geochim. Cosmochim. Acta 60, 3359–3360.
10. Craig, H., 1961. Isotopic variations in meteoric waters. Science 133, 1702–1703.
11. Dewandel, B., Lachassagne, P., Wyns, R., Maréchal, J.C., Krishnamurthy, N.S., 2006. A generalized hydrogeological conceptual model of granite aquifers controlled by single or multiphase weathering. J. Hydrol. 330, 260–284.

12. Dokka, R., 1983. Displacements on late Cenozoic strike-slip faults of the Central Mojave Desert, California. Geology 11, 305–308.
13. Dokka, R.K., 1986. Patterns and modes of early Miocene extension of the central Mojave Desert California. In: Mayer, L. (Ed.), Extensional Tectonics of the Southwestern United States, vol. 208. Geological Society of America Special Paper, pp. 75–95.
14. Dokka, R., Travis, C., 1990. Late Cenozoic strike-slip faulting in the Mojave Desert, California. Tectonics 9, 311–340.
15. Dokka, R.K., Henry, D.J., Ross, T.M., Baksi, A.K., Lambert, J., Travis, C.J., Jones, S.M., Jacobson, C.E., McCurry, M.M., Woodburne, M.O., Ford, J.P., 1991. Aspects of the Mesozoic and Cenozoic geologic evolution of the Mojave Desert. In: Walawender, M.J., Hanan, B.B. (Eds.), Geological Excursions in Southern California and Mexico. Geological Society of America Annual Meeting Field Trip Guidebook, San Diego, California.
16. Dokka, R., Macaluso, Y., 2001. Topographic effects of the Eastern California Shear Zone in the Mojave Desert. J. Geophys. Res. 106, 20644–30625.
17. ESA (Environmental Science Associates), 2004. Mojave Water Agency 2004 Regional Water Management Plan, SCH#: 2003101119, Program Environmental Impact Report.
18. Faulkner, D.R., Jackson, C.A.L., Lunn, R.J., Schlische, R.W., Shipton, Z.K., Wibberley, C.A.J., Withjack, M.O., 2010. A review of recent developments concerning the structure, mechanics and fluid properties of fault zones. J. Struct. Geol. 32, 1557–1575.
19. Fraser, D.C., 1969. Contouring of VLF-EM data. Geophysics, 34958–34967.
20. Garfunkel, Z., 1974. Model for the late Cenozoic tectonic history of the Mojave Desert, California and for its relation to adjacent areas. Geol. Soc. Am. Bull. 85, 1931–1944.
21. GSI (Geothermal Surveys, Inc.), 2000. Results of Detailed Gravimetric Investigation, Yucca Mesa Area, San Bernardino County, California. Prepared for High Desert Water District. June 12, 2000.
22. Gudmundsson, A., 2001. Fluid overpressure and flow in fault zones: field measurements and models. Tectonophysics 336, 183–197.
23. Haugerud, R., Harding, D., Johnson, S., Harless, J., Weaver, C., 2003. High-Resolution Lidar Topography of the Puget Lowland, 13. GSA Today, Washington, pp. 4–10.
24. Izbicki, J., 2004. Source and Movement of Ground Water in the Western Part of the Mojave Desert, Southern California, USA. USGS Water-Resources Investigations Report 03-4313.
25. Izbicki, J., Michel, T., 2004. Movement and Age of Ground Water in the Western Part of the Mojave Desert, Southern California, USA. USGS Water-Resources Investigations Report 03-4314.

26. Knipe, R., 1993. The influence of fault zone processes and diagenesis on fluid flow. In: Horbury, A., Robinson, A. (Eds.), Diagenesis and Basin Development, 36. American Association of Petroleum Geologists Studies in Geology, pp. 135–151.
27. Lehmann, K.K., Berden, G., Engeln, R., 2009. An introduction to cavity ring-down spectrometry. In: Berden, G., Engeln, R. (Eds.), Cavity Ring-Down Spectrometry Techniques and Applications. John Wiley & Sons Ltd., United Kingdom, pp. 1–26.
28. Lewis, R.E., 1972. Ground-water Resources of the Yucca Valley-Joshua Tree Area, San Bernardino County, California. USGS Open-File Report 72-234.
29. Londquist, C.J., Martin, P., 1991. Geohydrology and groundwater flow simulation of the Surprise Spring Basin Aquifer System, San Bernardino County, CA. USGS Water Resources Investigation Report 89-4099.
30. Lunn, R.J., Wilson, J.P., Shipton, Z.K., Moir, H., 2008. Simulating brittle fault growth from linkage of preexisting structures. J. Geophys. Res.–Solid Earth 113, B07403.
31. Meisling, K.E., (PhD thesis) 1984. Neotectonics of the North Frontal Fault System, of the San Bernardino Mountains, Southern California: Cajon Pass to Lucern Valley. California Institute of Technology.
32. MWA (Mojave Water Agency), 2005. Este Hydrologic Atlas.
33. MWA (Mojave Water Agency), 2007. Geologic map for a portion of the Eastern Mojave Desert, San Bernardino County, CA, 1: 250,000.
34. Murad, A.A., Krishnamurthy, R.V., 2008. Factors controlling stable oxygen, hydrogen and carbon isotope ratios in regional groundwater of the eastern United Arab Emirates (UAE). Hydrol. Process. 22, 1922–1931.
35. Palacky, G.J., Ritsema, I.L., Jong, S.J., 1981. Electromagnetic prospecting for groundwater in Precambrian Terrains in the Republic of Upper Volta. Geophys. Prospect. 29, 932–955.
36. Paterson, N.R., Ronka, V., 1971. Five years of survey with the very low frequency electromagnetic method. Geoexploration 9, 7–26.
37. Smith, G., Pimentel, I., 1998. Regional Water Table (1998) and Groundwater-Level Changes in the Mojave River and the Morongo Ground-Water Basins, San Bernardino County, California. USGS Water-Resources Investigation Report 00-4090.
38. Smith, G., Friedman, I., Gleason, J.D., Warden, A., 1992. Stable isotope composition of waters in southeastern California: 2. Groundwaters and their relation to modern precipitation. J. Geophys. Res. 97, 5813–5823.
39. Stamos, C.L., Predmore, S.K., 1995. Data and water-table map of the Mojave River ground-water basin, San Bernardino County, California, November 1992: U.S. Geological Survey Water-Resources Investigations Report 95-4148, scale 1:125,000.

40. Stamos, C., Martin, P., Nishikawa, T., Cox, B., 2001. Simulation of Ground-Water Flow in the Mojave River Basin, California. USGS Water-Resources Investigations Report 01-4002 Version 3.
41. Stamos, C., Cox, B., Izbicki, J., Mendez, G., 2003. Geologic Setting, Geohydrology and Ground-Water Quality near the Helendale Fault in the Mojave River Basin, San Bernardino Country, California. USGS Water-Resources Investigations Report 03-4069.
42. Sultan, M., Wagdy, A., Manocha, N., Sauck, W., Gelil, A., Youssef, A.F., Becker, R., El Alfy, Z., Jones, C., 2007. An integrated approach for identifying aquifers in transcurrent fault systems: the Najd shear system of the Arabian Nubian shield. J. Hydrol. 349, 475–488.
43. Telford, W.M., Geldart, L.P., Sheriff, R.E., 1990. Applied Geophysics, second edition. Cambridge University Press.
44. Trayler, C.R., Koczot, K.M., 1995. Regional Water Table (1994) and Water-Level Changes in the Morongo Basin, San Bernardino County, California. USGS Water-Resources Investigations Report 5-4209.
45. USGS (United States Geological Survey), 2004. Precipitation History of the Mojave Desert Region, 1893–2001. U.S. Department of the Interior, USGS Fact Sheet 117-03.
46. Vallèe, M., Chouteau, M., Palacky, G.J., 1992. Effect of temporal and spatial variations of the primary signal on VLF total-field surveys. Geophysics 57, 97–105.

CITATION

D. Dailey, W. Sauck, M. Sultan, A. Milewski, M. Ahmed, W.R. Laton, R. Elkadiri, J. Foster, C. Schmidt, T. Al Harbi, Geophysical, remote sensing, GIS, and isotopic applications for a better understanding of the structural controls on groundwater flow in the Mojave Desert, California, Journal of Hydrology: Regional Studies, Volume 3, March 2015, Pages 211-232, ISSN 2214-5818, http://dx.doi.org/10.1016/j.ejrh.2014.12.002.

CHAPTER 6

An Overview of Integrated Remote Sensing and Gis for Groundwater Mapping in Egypt

Salwa Farouk Elbeih

Engineering Applications and Water Division, National Authority for Remote Sensing and Space Sciences, 23 Joseph Tito st., El-Nozha El-Gedida, Cairo, Egypt

ABSTRACT

Groundwater is considered the major portion of the world's freshwater resources. One of the main challenges facing the sustainable development of Egypt is the need for better management of its limited fresh water resources. Groundwater exists in the Nile Valley, Nile Delta, Western Desert Oases, and Sinai Peninsula. Hydrogeological mapping of groundwater resources is one of the main tools for the controlled development of groundwater resources. Remotely sensed surface indicators of groundwater provide useful data where practical classical alternatives are not available. Integrated remote sensing and GIS are widely used in groundwater mapping. Locating potential groundwater targets is becoming more convenient, cost effective than invasive methods and efficient with the advent of a number of satellite imagery. The nature of remote sensing-based groundwater exploration is to delineate all possible features connected with localization of groundwater. Data, driven out of remote sensing, support decisions related to sustainable development and groundwater management.

INTRODUCTION

In one of their most recent reports, the United Nations Population Division projects stated that the Egyptian population is expected to reach 100,909 million by the year 2025 and 123,452 million in 2050 [1]. Most of the

Egyptian population (97%) lives in the Nile Valley and Delta, which represent about 4% of the territory [2].

Groundwater accounts for 26% of global renewable fresh water resources [3]. Salt water (mainly in oceans) represents about 97.2% of the global water resources with only 2.8% available as fresh water. Surface water represents about 2.2% out of the 2.8% and 0.6% as groundwater. The problem is not only to locate the groundwater, as it is often imagined, but the engineer's problem usually is to find water at such a depth, in such quantities, and of such quality that can be economically utilized. Groundwater is a term used to denote all the waters found beneath the ground surface [4]. Groundwater aquifers are not just a source of water supply, but also a vast storage facility providing great management flexibility at relatively affordable costs.

One of the main challenges facing the sustainable development in Egypt is the need for better development and management of its limited fresh water resources. Some of the most important groundwater problems in Egypt are: over-exploitation and lowering of its levels in some areas; saltwater intrusion into freshwater aquifers in some of the Nile Delta governorates; waterlogging and/or salinization.

With the advance in space technology, it is now possible to employ remote sensing techniques for estimating surface and subsurface water over large areas. These methods are very useful for rapid groundwater mapping of large and inaccessible areas. The necessity of remote sensing based groundwater exploration is to demarcate and delineate all possible features connected with localization of groundwater. These features are extracted from the appropriate satellite data products and integrated with the thematic details obtained from topographic sheets of the desired scale. The use of conventional techniques, (e.g., geophysical, statistical or geostatistical techniques, numerical modeling, etc.) for groundwater management, is often severely limited by the lack of adequate data. Frequent and long-term monitoring of groundwater and vadose zone systems by these conventional methods is expensive, laborious, time-consuming and destructive [5]. Innovative technologies, as remote sensing (RS) and Geographic Information Systems (GIS), have an immense role to play [5].

WATER RESOURCES IN EGYPT

Egypt lies in the northeastern corner of the African Continent and has a total area of about 1 million km^2. The Egyptian terrain consists of a vast desert plateau dissected by the Nile Valley and its Delta, which occupy about 4% of the total country area. Since ancient times the Nile has been the main source of fresh water to the country covering all water needs for Egypt's population, which inhabited the Nile Valley and the Delta. Human activities in the remaining 97% desert land remained confined to a few localities, where deep groundwater was available through springs and seepage zones (valleys, oases, and depressions) [6]. Under the 1959 Nile Waters Agreement between Egypt and Sudan, Egypt's share of the Nile flow is 55.5 billion m^3/year. Egypt is a water scarce country, with less than 1000 m^3 of available freshwater/year/capita. The rate of water usage in Egypt is approximately 190 L/person/day. On the other hand, water needs for typical Middle Eastern-North African diet is 294 L/person/day. The annual average rainfall is 51 mm [7].

Table 1 shows the water budget of Egypt in 2010; as given in the 2050 water strategy of the Ministry of Water Resources and Irrigation [7] and [8].

Table 1. Current water budget of Egypt (2010), and all sources and allocation/usage [7].

2010 Water budget (sources and usage), population = 80 millions				
Water Supply	Volume (billion m^3/year)	Demand by sector	Consumption (billion m^3/year)	Usage/allocation (billion m^3/year)
Conventional water sources		Drinking (Fresh W only)	1.8	9
Nile (HAD)	55.5	Industry	1.4	2
Deep groundwater	2	Agriculture	40.4	67
Rainfall and flash floods	1.3	Drainage to sea	12.2	
Desalination	0.2	Evaporation losses	3	3
Total supply conventional	59	Env. balance	0.2	0.2
Unconventional sources		Total consumption	59	
Shallow groundwater (Delta)	6.2			
Re-use of Agricultural drainage water	16			
Total supply non-conventional	22.2			
Total water supply	81.2	Total water usage or allocation		81.2

From this water budget, it is clear that Egypt's water demand for irrigation, industry, and domestic consumption already exceeds the supply of the Nile. This is substituted by recycling fresh water more than once, which imply that there is shortage in the fresh water resources, and also reflects the high efficiency of the system as well as the sensitivity of the system to deterioration in water quality problems that may arise [7].

Groundwater exists in the fringes of the Nile Valley, Nile Delta, Western Desert, and Sinai Peninsula. Groundwater is the only available resource for interdisciplinary development in the areas around the Nile Valley, because of insufficient rainfall. Groundwater in the Nile aquifer system and desert fringes is not a resource in itself as it is replenished from the river Nile by seepage from canals and deep percolation from irrigation application [9]. Groundwater in the Western Desert is deep seated. Recent studies have indicated that this is not a renewable resource and use of this fossil water depends on the cost of pumping and potential economic return over a fixed time period [10].

On the other hand, in the coastal zones, groundwater reservoirs are recharged from local rainfall [2]. The largest groundwater deposit is the giant Nubian sandstone aquifer underneath the eastern part of the African Sahara, and is shared between Egypt and four other countries [11]. The hydrogeological framework of Egypt comprises six aquifer systems [12], [2], [13], [10], [14], [6] and [9], as shown in Fig. 1:

(1) *The Nile aquifer system*, assigned to the Quaternary and Late Tertiary, occupies the Nile flood plain and desert fringes. The aquifer consists of a thick layer of graded sand and gravel covered by a clay layer in its major part. The aquifer thickness varies from 300 m (at Sohag) to only a few meters (at Cairo and Aswan) North of Cairo. The aquifer system is renewable and the main recharge source is the infiltration from the excess water application for agriculture and seepage from the irrigation water distribution system.
(2) *The Nubian Sandstone aquifer system*, assigned to the Paleozoic–Mesozoic, occupies a large area in the Western Desert, and parts of the Eastern Desert and Sinai. Groundwater can be found at very shallow depths, where the water bearing horizon is exposed; or at very large depths (up to 1500), where the aquifer is semi-confined. The Nubian Sandstone aquifer system is regional system. It extends into Libya, Sudan and Chad and it is non-renewable aquifer system [15].
(3) *The Moghra aquifer system*, assigned to the Lower Miocene, occupies mainly the western edge of the Delta up to the Qattara depression. The Moghra outcrops on the surface in Wadi el Natrun and Wadi el Farigh. Recharge is achieved mainly though deep percolation from the Nile

alluvium and groundwater flow across the common boundary of the Moghra and the Nile alluvium [16].

(4) *The Coastal aquifer systems*, assigned to the Quaternary and Late Tertiary. They are found in the littoral zones long the Mediterranean and Red seacoasts in the form of scattered pockets. Groundwater is generally under phreatic conditions in the form of thin lenses floating over saline water. The aquifer is renewable and it is replenished from rainfall and exploited large numbers of shallow wells and galleries[17].

(5) *The karstified carbonate aquifer system*, assigned to the Eocene and to the Upper Cretaceous, predominates essentially in the north and middle parts of the Western Desert. The fissured and karstified carbonate aquifer complex is the least explored and exploited in Egypt, although it occupies at least 50% of the total area of the country. The water recharge to the aquifer depends essentially on the upward leakage from the underlying Nubian Sandstone aquifer and occasionally from local rainfall.

(6) *The fissured and weathered hard rock aquifer system*, assigned to the Pre-Cambrian, predominates in the Eastern Desert and Sinai. This aquifer is influenced by several factors, namely tectonic, morpho-tectonic and lithologic. The tectonic factors such as filled fractures and dykes may hinder the movement of groundwater.

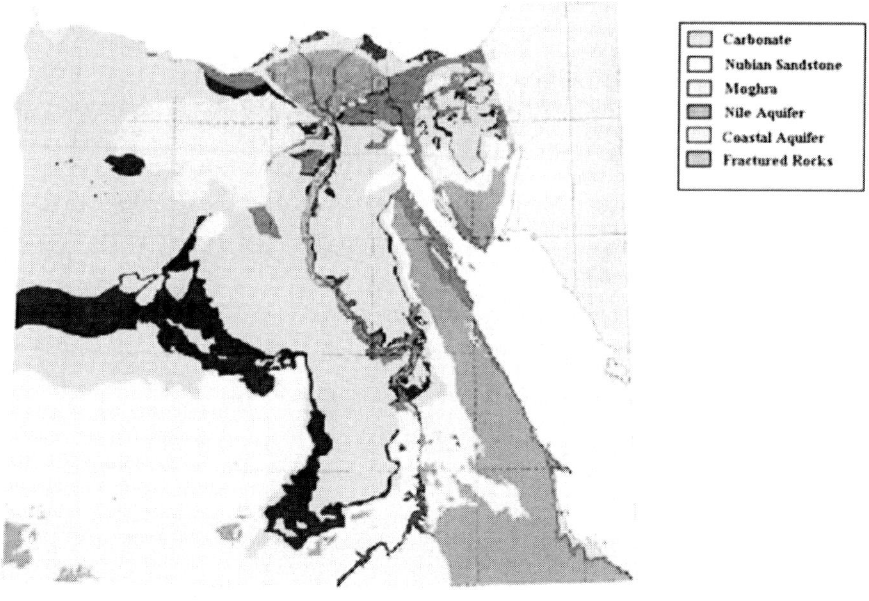

Figure 1. Aquifer systems in Egypt [12] and [13].

GROUNDWATER MAPPING

Concept and benefits of groundwater mapping

Groundwater (hydrogeological) mapping is one of the main tools for systematic and controlled development and planning of groundwater resources. These maps are used by engineers, planners and decision makers in order to allocate, develop and manage groundwater within a national water policy [18]. Hydrogeological maps present hydrogeological data in a map form. A hydrogeological map shows the geographical distribution of aquifers, and their topographical, geological, hydrographical, hydrological and hydrochemical features. Presentation of these data in the form of maps permits the rapid evaluation of a certain area. Accordingly, hydrogeological maps assist in determining areas needing special protection [18].

In 1985, the Research Institute for Groundwater (RIGW), under the National Water Research Center (NWRC) of the Ministry of Water Resources and Irrigation published a series of hydrogeological maps of Egypt. These maps were comprised of maps at different scales; detailed maps (1:100,000) for the Nile Valley, Nile Delta, Oasis and Coastal areas; semi-detailed maps (1:500,000), and a regional map (1:2,000,000) for Egypt [13].

Table 2 shows the working procedure in the compilation of the 1:100,000 and 1:500,000 hydrogeological maps. The procedure involves: Data collection, processing, interpretation and printing. The map legend is based on the second edition of the "International legend for hydrogeological maps" published by UNESCO [19].

Table 2. Working procedure for the preparation of a hydrogeological map.

Step	Description
Data collection	• Collection of existing data • Field investigations to complete data • Preparation of remote sensing maps
Data processing	Preparation of basic maps • Topography • Surface water • Well data • Groundwater extraction • Geology
Data	Preparation of interpretation maps

interpretation	• Base of aquifer • Piezometric contours • Salinity contours • Aquifer units Preparation of inset maps and cross sections Compilation of hydrogeological maps: join basic and interpretation map to form one map
Printing	• Transferring data to digital format, using the GIS ARC/INFO • Printing of maps • Preparation of explanatory notes

Source: Khalil and Mebius [18]

The hydrogeological map of Egypt (1:2,000,000) is mainly used for delineating regional aquifers, selecting potential areas for regional groundwater development. Promising areas selected are the Nubian Sandstone basin, large inland wadi catchments (Wadi Qena and Wadi El-Assyuti and others) and the littoral zones. The sub-regional maps (scale 1:500,000) are used to study the feasibility of these groundwater resources for development. In addition, the wide distribution of the 1:2,000,000 map has led to a growing awareness among decision makers on the importance of groundwater and the need for careful planning. The detailed maps (1:100,000) are used in different applications; rising water table, influence of groundwater table on monuments, assessment of groundwater vulnerability to pollution [20].

Elements of a groundwater map

The first hydrogeological maps of Egypt were produced based on the available field data and reports and were hand drawn. The 2nd generation maps were produced by digitizing available data, including topographic maps, landuse maps, geological maps and surface water maps.

As an example for a hydrogeological map elements, the nine layers of the 1:500,000 scale hydrogeological Map of the Egyptian South Western Desert are as follows[20] and [21].

1 – Hydrogeological units map

- Intergranular Aquifers.
- Rocks With Limited Groundwater Resources.

- Karstified and Fissured Aquifers.
- Non-Aquifers (Aquiclude).

2 – Lithological map

Data of the lithological map are extracted from geological maps, satellite images, hydrogeological units and the main tectonic features.

A. Quaternary.
B. Tertiary.
C. Mesozoic.
D. Mesozoic–Paleozoic.

3 – Surface water map

Produced from topographic maps and/or satellite images. The main inputs for the surface water map are:

a) Perennial stream with direction of flow and average annual water level relative to mean sea level (MSL).
b) Intermittent stream (Wadi).
c) Secondary surface water divide.
d) Water logged area.
e) Playa depression acting as a temporary accumulation area for surface runoff water.
f) Drains and canals.

4 – Man made features map

A. Groundwater Extraction: It represents average discharge in m^3/day.
B. Main irrigation canal.
C. Barrage.
D. Surface water pumping station.
E. Pipeline for water supply from surface water.

The sources of these data include: Data base from Research Institute of Groundwater (RIGW), reports and field measurements.

5 – Groundwater features

a) Wells and Boreholes.
b) Observation well with identification number.
c) Production well with identification number.

d) Group of production wells.
e) Hand dug well with identification name.
f) Spring.
g) Vertical Electrical Sounding (VES) with identification number.

The sources of these data include: Data base from Research Institute of Groundwater (RIGW), reports and field measurements.

6 – Topographic map

Produced from colored topographic maps published by the Egyptian Survey and/or SRTM data. It includes the following information:

a) Topographical contour line, elevation in meter relative to mean sea level.
b) Triangulation point.
c) Main town.
d) Main road.
e) Nile Valley edge.
f) Boundary of new reclaimed area.
g) Boundary of old cultivated land.
h) Water divide.

7 – Base of Aquifer map

Base of Quaternary Aquifer relative to mean sea level.

a) Base of Quaternary Aquifer in m relative to mean Sea level.
b) Base of Nubian Sandstone aquifer in m relative to mean Sea level.
c) Base of Carbonate aquifer.

The sources of these data include: Data base from Research Institute of Groundwater (RIGW), geological maps and reports

8 – Isopiezometric map

a) Contour line of the average piezometric head of the Quaternary Aquifer in m relative to mean sea level.
b) Contour line of the average piezometric head of the Nubian sandstone aquifer.
c) Contour line of Piezometric head of fissured carbonate aquifer in m relative to mean sea level.
d) Direction of Groundwater flow.

9 – Isosalinity map

a) –Isosalinity line of the Quaternary Aquifer, dashed where uncertain.

b) –Isosalinity line of the Nubian sandstone aquifer.

RS AND GIS FOR GROUNDWATER MAPPING

Applications of RS and GIS in groundwater mapping

General

Groundwater specialists were late to use satellite data for an obvious reason; groundwater lies in the subsurface. Moreover, the current air- and satellite-based radar and radiometers can normally penetrate only few centimeters into the ground. In spite of this apparent obstacle, RS holds tremendous potential for regional groundwater flow studies [22]. Remotely sensed indicators of groundwater may provide important data where classical alternatives are not available. Examples of measured data include groundwater heads, changes in groundwater storage, heat signatures, and subsidence data [22]. Ground-based RS (geophysics) is usually more expensive than space and airborne RS but is still more accurate and cheaper than invasive methods (boring drilling)[23]. These indicators include vegetation surface water, water discharging to the surface carrying heat energy and runoff [22]. Satellite technology is reviewed with respect to its ability to measure groundwater potential, storage, and fluxes. Satellite data (Table 3) can be used if ancillary analysis is used to infer groundwater behavior from surface expressions. Remotely sensed data are most useful where they are combined with numerical modeling, GIS, and ground-based information [22].

A key to the RS of groundwater is the recognition that shallow groundwater flow is usually driven by surface forcing and parameterized by geologic properties that can be inferred from surface data. A particularly successful application of RS to groundwater has been the identification of lineaments that are thought to be related to faults and fracturing in hard-rock [22]. A straightforward way to couple RS with

Table 3. A selected list of active space-based sensors that report data of potential use for investigations of groundwater [22].

Sensor	Launch year	Ground resolution (m)	Precipitation	Surface temperature	Soil moisture	Water storage	Snow water	Land cover	Topography
AMSR-E	2002	5400–56,000	x	x	x		x		
ASTER	1999	15,30,90		X				X	X
AVHRR	1991–2003	1100	X	X	X		X		
GRACE	2002	300,000				X			
ENVISAT-RA[2]	2002	1000		x					X
Landsat-7	1999	30,60		X				X	
MODIS	1999	250,50,1000		X				X	
OrbView-2	1997	1100		x				X	
OrbView-3	2003	1,4		X				X	X
RADARSAT-1	1995	8–100			X				
SRTM	2000	30,90							X

groundwater flow predictions, therefore, is to use remotely sensed imagery to define boundary conditions such as streams, lakes, wetlands, seepage areas, recharge zones, or evapotranspiration zones.

Jasmin and Mallikarjuna [24] presented the application of RS and GIS in the exploration and assessment of groundwater in consolidated and unconsolidated formations in semi-arid regions, particularly with reference to India. They also highlighted the importance of GIS technology in providing an efficient handling and management of large and complex databases for groundwater assessment studies. The use of high-quality remotely sensed data, identification of pertinent hydrological, geological and geomorphological parameters, preparation of corresponding thematic maps, assigning appropriate weights and integration in a sophisticated GIS environment ensure and enhance the prediction accuracy of location of promising zones and their groundwater potential.

Knowledge of groundwater location is important for both water supply and pollution control analysis. The identification of topographic and vegetation indicators of groundwater and the determination of the groundwater location discharge area (seeps and springs) can assist in the location of potential well sites. Groundwater recharge zones could be identified in order to protect these areas (through zoning restrictions) from activities that would pollute the groundwater supply. Available image interpretation cannot be directly used to map the depth to water in a groundwater system. However, natural vegetation types are used as indicators of approximate depth to groundwater[25].

Visual image interpretation can be used in a different way to help in monitoring the quality, quantity and geographic distribution of water resources. In this concern, vegetation types are successfully used as an indication of approximate depth to groundwater [26].

Zevenbergen et al. [27] reported that integrated RS and GIS represent an additional source of information for hydrogeological maps production (Fig. 2) where the application of RS in the field of water resources management includes:

a) Hydrogeological mapping.
b) Identification of areas affected by salinization.
c) Determination of Biomass.
d) Land use and crop classification.
e) Environmental monitoring.

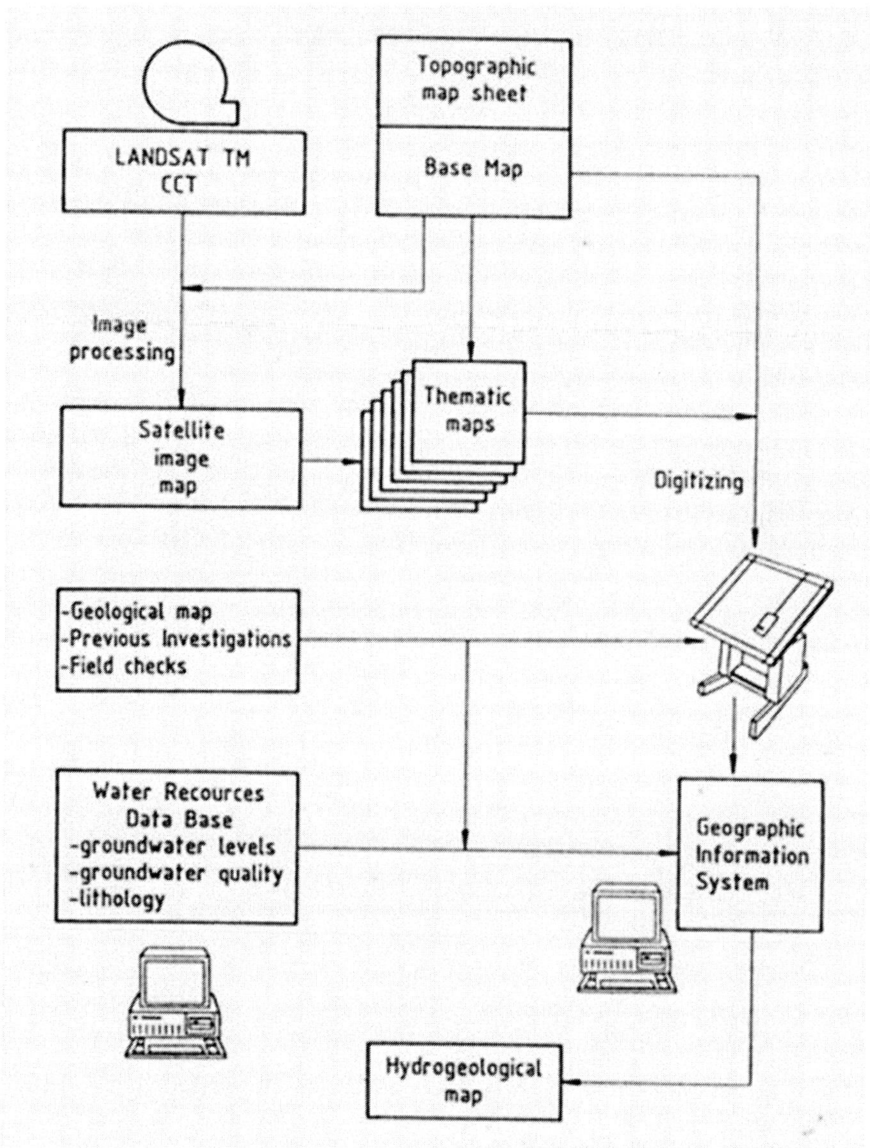

Figure 2. Procedure for production of satellite image maps and hydrogeological maps using RS and GIS [27].

Usage of aerial photographs and satellite images

Kwarteng [28] used aerial photographs, Landsat Thematic Mapper (TM) images and Digital Elevation Models (DEM) for mapping paleodrainage patterns, large depressions, playas and catchments areas. These parameters are used for accumulating large amounts of water during flash floods and recharging of freshwater lenses. These data sets were analyzed using GIS to delineate the most favorable areas for the formation of fresh groundwater lenses.

Jaiswal et al. [29] planned for groundwater exploitation by integrating a number of thematic maps (Fig. 3) in a GIS environment for depicting village-wise groundwater prospect zones in the Gorna sub-basin of Madhya Pradesh (Central India). Thematic layers of lithology, landforms, soils, land use/land cover, lineaments and surface water bodies were prepared using the Indian Remote Sensing (IRS) data, whereas, drainage density maps and slope classes were prepared from topographic maps. The feature-based categorizations of different thematic maps and the exact weight were assigned to each of the thematic features based on the relative merit and demerit with respect to groundwater occurrence. The authors adopted the following order of sequence to derive the final integrated layer.

Thematic/integrated layer	Resultant layer
Lithology (R1) + Landform (R2)=	S1
S1 + soil type (R3)=	S2
S2 + slope (R4)=	S3
S3 + drainage density (R5)=	S4
S4 + land use/land cover (R6)=	S5
S5 + lineament buffer (R7)=	S6
S6 × surface water bodies layer (R8)=	S7

Source: Jaiswal et al. [29].

From the composite layer (S7), the delineation of groundwater prospect zones was made by grouping the polygons into different prospect zones: excellent, good, moderate and poor.

Oh et al. [30] presented a probabilistic approach that used both satellite imagery and GIS to estimate an area's potential groundwater resources. The adopted procedure involves selecting the 15 most important variables that affect the groundwater potential. The groundwater potential was then mapped using a frequency-ratio model, which represents the relationship between hydrologic data; specific capacity (derived from pumping tests) and the variables. In particular, 9 of the 15 factors showed a more positive influence than the remaining factors on groundwater potential mapping. In an increasing order of positive influence, these nine factors were soil texture, cumulative watershed area, distance from river, mean ground slope within watershed area, lineament length density weighted by its length, lineament length density, mean ground elevation within watershed area, topographic wetness index (TWI), and hydrological units.

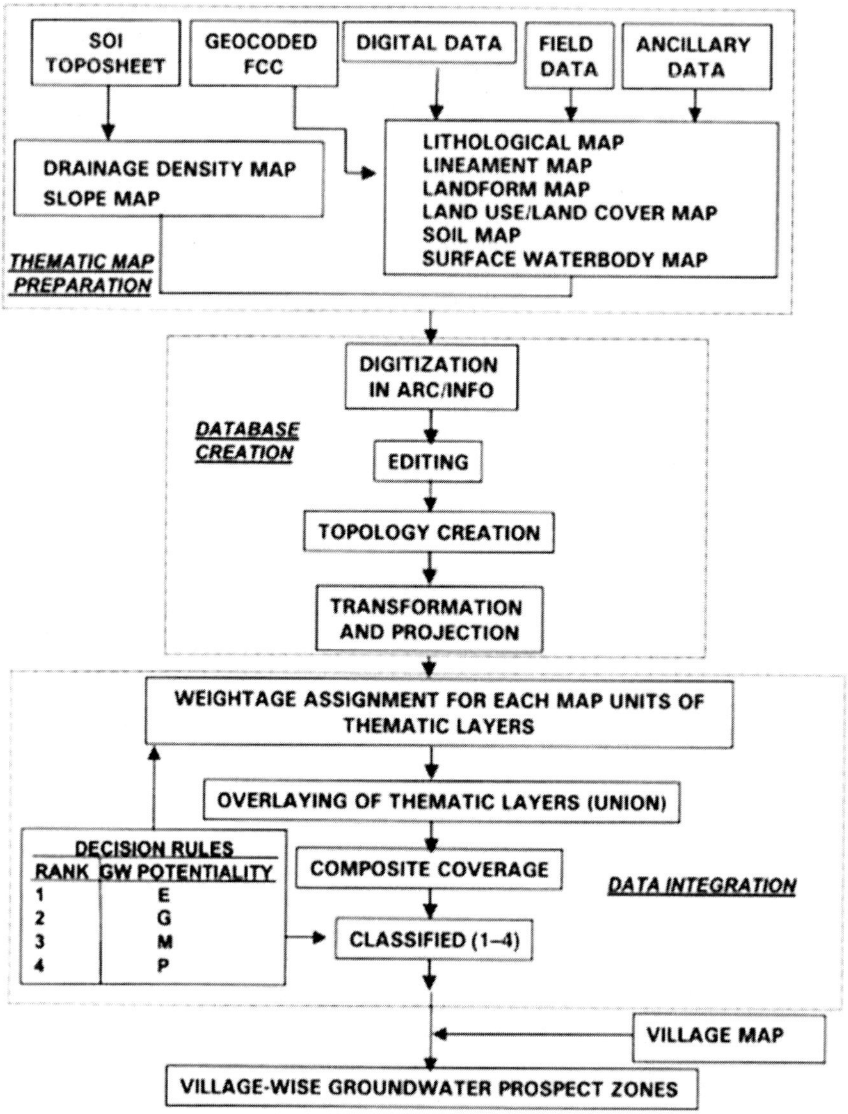

Figure 3. Approach for delineating groundwater prospect zones.

Droogers et al. [31] described the recent data from the GRACE (Gravity Recovery and Climate Experiment) Satellite, available to assess changes in terrestrial water over larger areas. GRACE is a twin-satellite mission used to identify mass changes due to variations in water storage, assist in determination of groundwater depletion, residual basin-scale estimates of evaporation or validation of hydrological models. GRACE consists of two polar orbiting satellites; developed to fly at an altitude ranging from 300 to 500 km and separated by a distance of about 200 km along track. The earth's gravity field causes accelerations of the satellites where they approach an area of relatively high mass concentration, and decelerations where they move away from them (Fig. 4). GRACE data are expressed in mm equivalent to water but with no distinction between snow cover, soil moisture and deep groundwater storage.

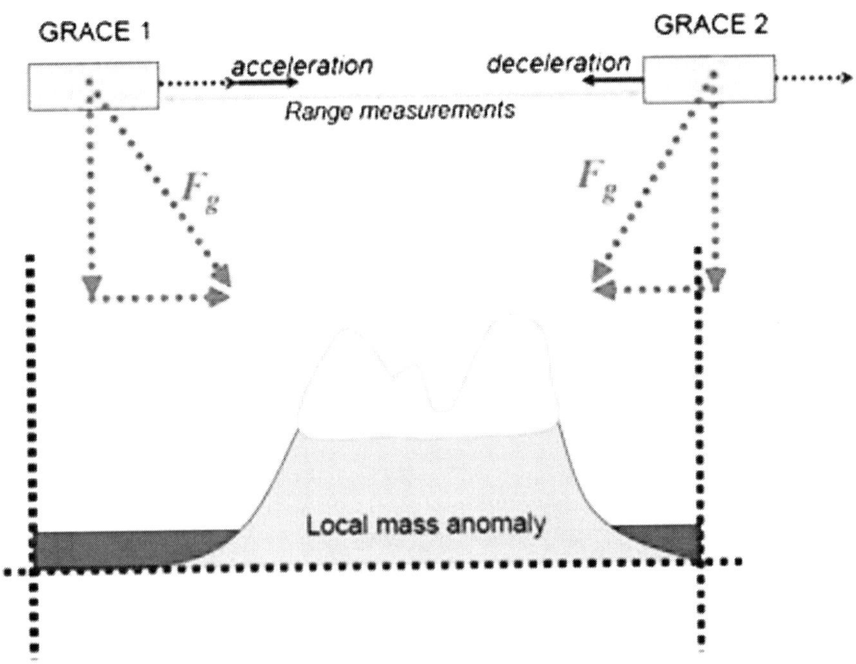

Figure 4. A schematic cartoon of the way in which GRACE measures the gravity field [31].

Usage of radar data

Robinson et al. [32] used seven Radarsat-1 data scenes to delineate the drainage of two wadi systems in Kufra Oasis, Libya. These wadis drain into the Kufra Oasis in southeastern Libya from the south and southeast. Drainage features are digitized on the individual scenes and are mosaicked. The interpretation of the Radarsat-1 scenes allows the definition of passageways and intermediate holding areas of the courses of former rivers. These results offer some explanation for the amount of water that has been retrieved from the Kufra wells for over four decades.

According to Meijerink [23], microwave or radar images have several applications in hydrogeological applications; Geological structures and lineaments, Dynamics of large marshes (always connected to groundwater and can be analyzed using radar images). Radar altimetry for detecting lake levels is important for the groundwater heads and flow studies around lakes. Moreover, Digital Elevation Models (DEMs) and precise measurement of land subsidence are achieved via radar data. Phase shifts of two or more radar images of the same area make it possible to determine, with high precision, changes in land elevation. Land subsidence, often related to pumping or recharge of aquifers, can also be mapped [23].

GROUNDWATER MAPPING IN EGYPT USING RS AND GIS (CASE STUDIES)

Eastern Desert

Omran [33] performed an interpretation approach that integrates data obtained from RS, geophysics and other hydrogeologic phenomena through GIS, to characterize groundwater resources for the identification of candidate well locations in Sohag region. Groundwater potential zones have been demarcated by integration of aquifer thickness derived from surface electrical resistivity survey and drilling data with the different thematic layers; hydrogeomorphology, lineaments, slope, drainage and depth to aquifer, using a model developed through a GIS technique. The weights for each parameter are shown in Table 4. The groundwater potential zone map generated (Fig. 5) through this model was verified with the yield data to ascertain the validity of the model developed. Total weights of final integrated grids were derived as a sum of the weights assigned to the different layers based on suitability.

Table 4. Weightage of different parameter for groundwater prospects [33].

S. no.	Criteria	Classes	Weight
1	Hydrogeomorphology	Nile flood plain	5
		Transverse channels	3
		Low land desert areas	2
		Limestone plateau	1
2	Slope (degree)	0–0.5	5
		0.6–2.0	4
		2.1–5.0	3
		5.1–10.0	2
		>10.0	1
3	Surface lineament	Present	2
		Absent	1
4	Subsurface lineament	Present	2
		Absent	1
5	Drainage	High possibility for floods (A)	3
		Moderate possibility for floods (B)	2
		Low possibility for floods (C)	1
6	Depth to water bearing zone	>25 m	3
		6.0–25.0 m	2
		<6.0 m	1
7	Aquifer thickness	>35 m	5
		26.0–35.0 m	4
		16.0–25.0 m	3
		6.0–15.0 m	2
		<6.0 m	1

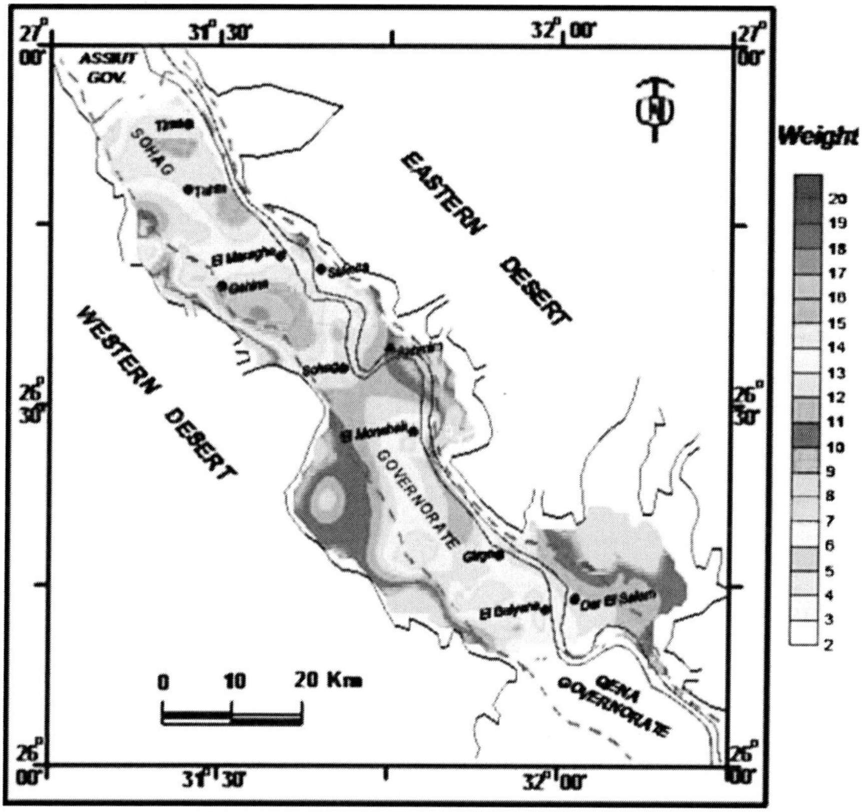

Figure 5. Groundwater potentiality map, Quaternary Aquifer, Sohag Governorate area [33].

Abdalla [34] determined the most important contributing parameters that indicate the groundwater potential in Wadi El Laqeita in the Central Eastern Desert and its surroundings as one of the most promising areas for agricultural development in Egypt. To delineate the location of the groundwater prospect zones, different thematic maps were prepared from topographic and geological maps. These maps were based on determining the most important contributing parameters that control groundwater storage. These parameters include the following:

(1) slope, which controls the runoff of water or remains on the ground surface for time long enough to infiltrate (Sl),
(2) stream network, which influences the distribution of runoff and groundwater recharge (Dr),

(3) lineament, which enhance significantly the permeability by inducing secondary porosity and hence vertical water percolation to recharge the aquifers (Lin),
(4) lithology or rock type, which determines the soil and exposed rocks infiltration capabilities and govern the flow and storage of water (Geo) and
(5) topographical map layers (top).

The thematic map of each parameter was produced using GIS and RS. These input layers were combined using the Raster Calculate Module in a GIS model to produce the final groundwater prospective zones of the investigated area (Fig. 6) [34]. In this model each thematic map has been given a weight value depending on its strength of influence/contribution with respect to the groundwater storing (Table 5).

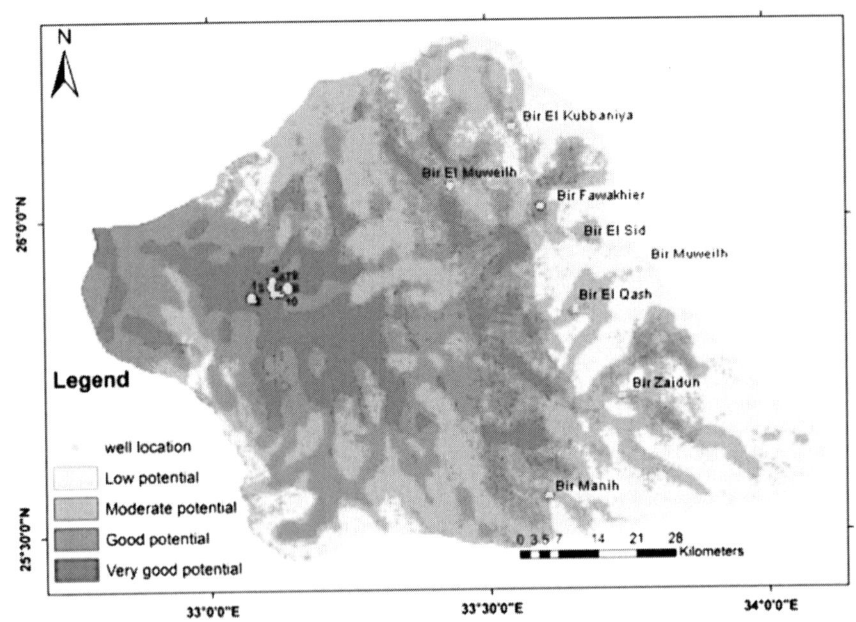

Figure 6. Groundwater prospect map of the Central Eastern Desert [34].

Table 5. Thematic map weight and capability value for groundwater prospective zones in the Central Eastern Desert [34].

Thematic layer	Map rank	Map weight (Wi)	Class ranges	Degree	Rank	Capability value
Slope	2	2/13 = 0.15	0–5	High	3	0.6
			5–10	Moderate	2	0.33
			10–44.041798	Low	1	0.16
Stream networks	3	3/13 = 0.23	0–0.30	Low	1	0.16
			0.30–0.50	Moderate	2	0.33
			0.50–0.848	High	3	0.5
Lithology	2	2/13 = 0.15	Basement	Low	1	0.1
			Sedimentary sequence	Moderate	2	0.2
			Nubian sandstone	High	3	0.3
			Wadi deposit	Very high	4	0.4
Lineament	3	3/13 = 0.23	0–20	Low	1	0.16
			20–40	Moderate	2	0.33
			40–0.62997	High	3	0.5
Topography	3	3/13 = 0.23	67–270	High	3	0.5
			270–450	Moderate	2	0.33
			450–1047	Low	1	0.16

The groundwater prospect map (Fig. 6) was mathematically calculated using ARC GIS raster analysis as follows:

$$GWP=\sum Wi*CVi$$

Where the GWP = groundwater potential, Wi = map weight, CVi = capability value (weight of inter-map class).

$$GWP=\sum(Dr,Lin,SI,Geo,top)$$

where Dr = stream network classes, Lin = lineaments density classes, Sl = slope classes Geo = lithology classes, top = topography/elevation classes.

Gheith and Sultan [35] constructed an integrated hydrology model that represents a first order estimation for groundwater from sparse storms events over the Red Sea hills. Watersheds and stream networks were identified from digital terrain models verified by comparison with co-registered Landsat scenes and geologic maps.

Amer et al. [36] assessed quality, quantity and potential uses for groundwater in the Central Eastern Desert of Egypt as a case study for groundwater utilization in arid regions. Field studies and RS data were combined with a GIS-based modeling to estimate the groundwater budget of sedimentary sub-basins, and use these estimates to suggest locations for new wells. Stable isotopic composition and major ion hydrochemistry of groundwater were used to evaluate the source of water in different aquifer types and propose management policies for groundwater use.

Sinai Peninsula

In 1988, a joint project between the Government of Arab Republic of Egypt and the United Nations Development Program (UNDP) and UNESCO was launched under the title: "Capacity Building of The Egyptian Geological and Mining Authority (EGSMA) and The National Authority for Remote Sensing and Space Sciences (NARSS) for the sustainable development of the South valley and Sinai". One of the main outputs of this joint project was the production of an atlas of water resources in Sinai Peninsula with a scale of 1:450,000 [21].

In 1992, the Japan International Cooperation Agency (JICA) published a hydrogeological map for North Sinai and in 1998 published a hydrogeological map of South Sinai with a scale of 1:250,000 [37] and [38].

Elewa and Qaddah [39] integrated Enhanced Thematic Mapper Plus (ETM+) images, GIS, a watershed modeling system (WMS) and weighted spatial probability modeling (WSPM) to identify the groundwater potential areas in the Sinai Peninsula, Egypt. Validation using measured well yield data was performed to check the WSPM results. Eight related thematic layers (rainfall amount, net groundwater recharge, lithology, lineament density, terrain slope, drainage density, depth to water, and water quality) were built in a GIS and assigned appropriate rankings. The WSPM was checked and validated by comparison with the published hydrogeological map of North Sinai in 1992 and actual borehole yield data and it was found that it correlates well with the previously published data and maps. The resulting groundwater potentiality map of Sinai (Fig. 7) indicates that the different geographic locations are suitable for groundwater storage with different magnitudes and potentialities, but the overall groundwater potential is of the moderate class.

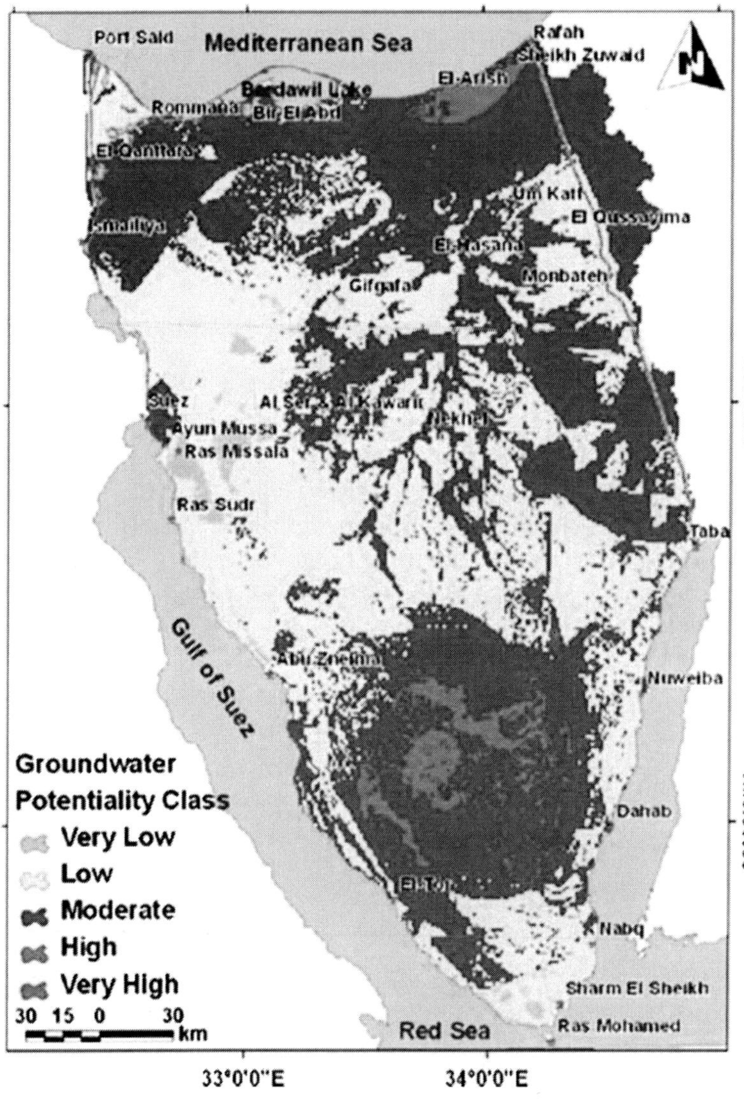

Figure 7. Groundwater potentiality map of Sinai [39].

Sultan et al. [40] integrated optical and radar RS and geophysical data for exploring groundwater in El Qaa Plain, South Sinai, Egypt. The hydrological analyses of the Digital Elevation Model (DEM) extracted the drainage networks pattern, determined the flow direction of the main channels and find the relationship between paleochannels and structural elements. The water table maps constructed from interpreted geoelectrical data indicates the elevation of the water table decreases from the east and north towards the west.

Milewski et al. [41] developed an interactive data language (IDL)-based computer program, the RS data extraction model (RESDEM) for integrated processing and analysis of a suite of RS data sets. RESDEM imports, calibrates, and georeference scenes, and subsets global data sets for the purpose of extracting and verifying precipitation over areas and time periods of interest. Case studies were conducted in Sinai Peninsula and the Eastern Desert.

Western Desert

Robinson [42] used SIR-C and Radarsat radar data to map structural and fluvial features in the southwest of Egypt to identify new groundwater resources in fractured rock aquifer settings. The interpretation of one Radarsat image and five high-resolution SIR-C radar scenes combined with Digital Elevation Models (DEM) analysis showed that NE and ENE drainage directions prevail in south-western Egypt and culminate in the Kharga Depression. Radar data are ideal for mapping palaeochannels since some are sand covered (L-band has the ability to penetrate to at least 2 m beneath the sand in arid areas). The regionally extensive faults (Fig. 8) drain numerous wadis and extend for tens of kilometers in length, so may constitute important water carrying conduits.

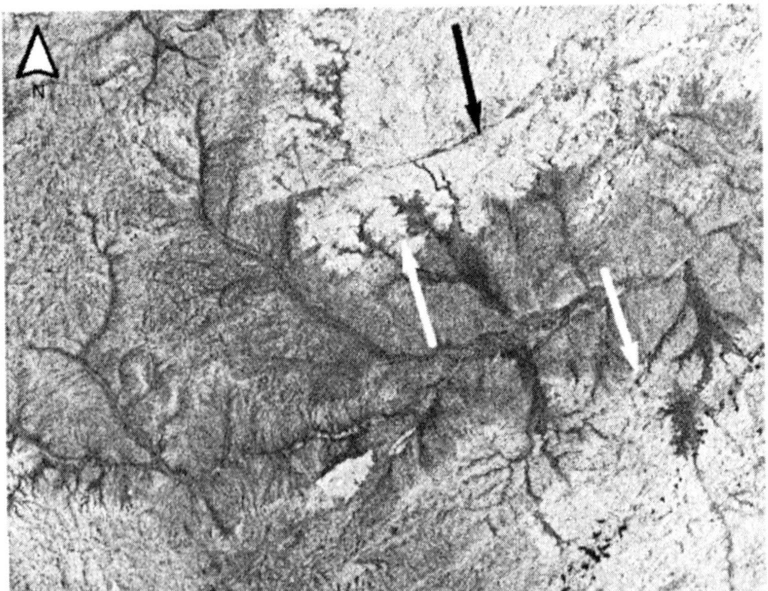

Figure 8. Part of the Radarsat scene showing well-developed trellis-like channels with pronounced structural control, superimposed on an older, structurally-controlled flood feature. N.B.: The white arrows point to the edges of the structurally controlled fluvial features while the black arrow points to a parallel fault in the adjacent bedrock [42].

Robinson [43] used C-band Radarsat-1 SAR images and topographic data to understand groundwater distributions in the Nubian aquifer, which underlies East Oweinat and Tushka regions. In this area, groundwater represents the best locations for groundwater development in order to address the water and food pressures that exist in these regions. Radar waves are uniquely able to penetrate beneath the desert sand in the eastern Sahara to reveal groundwater related near surface features, that is; courses of ancient rivers, streams, faults and fractures. The depth of near surface imaging for the data sets used is on the order of half a meter. Three new areas have been identified as promising sites for agricultural development.

Koch et al. [44] used optical and radar RS techniques, supported by limited ground surveys to locate potential areas for accumulating groundwater in El-Gallaba plain, west of Aswan City. In this area, the plain once hosted an ancestral river system long before the Nile even existed but nowadays the fluvial deposits are largely covered by Aeolian and gravelly sands and thus only detectible with radar and thermal images. These techniques were a combination of multispectral, thermal and microwave data (ASTER and PALSAR) and supported by ground measurements (Ground Penetration Radar (GPR), field spectroradiometry and magnetometry).

The infrared bands of ASTER image were used to calculate the Land Surface Temperature (LST) in El-Gallaba plain where the resulting LST map revealed two broad strips of thermal cooling anomalies arranged in a linear fashion and diagonally crossing the alluvial basin of the study area. The areas were further explored with radar data to better understand the nature of these LST anomalies. Radar ALOS/PALSAR data (HH) were used to determine the range of backscatter coefficient values of the sediments along the thermal anomalies and their surroundings. The mean backscatter coefficient (σ°) of the sediments along these linear thermal anomalies is −20 dB, while the mean value of the surrounding materials is −13 dB. This variation in backscatter coefficient suggests that the linear cooling anomalies have a grain size of sand and less than the surrounding materials which have a grain size ranging from coarse sand to gravel fractions [44].

A number of 12 GPR profiles across the LST anomalies confirmed that the near-surface sediments (up to 17 m depth) consist of thin horizontal layers of sandstone with very low gravel content. A set of spectral signatures were acquired at the investigated sites (LST anomalies sites and surroundings) where the results indicated that the silt and clay content is low as they generally do not exceed 1%.

A number of 12 magnetic profiles with 1.5 km average length were acquired across the linear LST anomalies to investigate deep-seated structures where the processed magnetic profiles confirmed the existence of graben-like structures with a maximum depth to basement of 160 m (Fig. 9) in the middle section and shallower depths toward the edges of the linear LST anomalies.

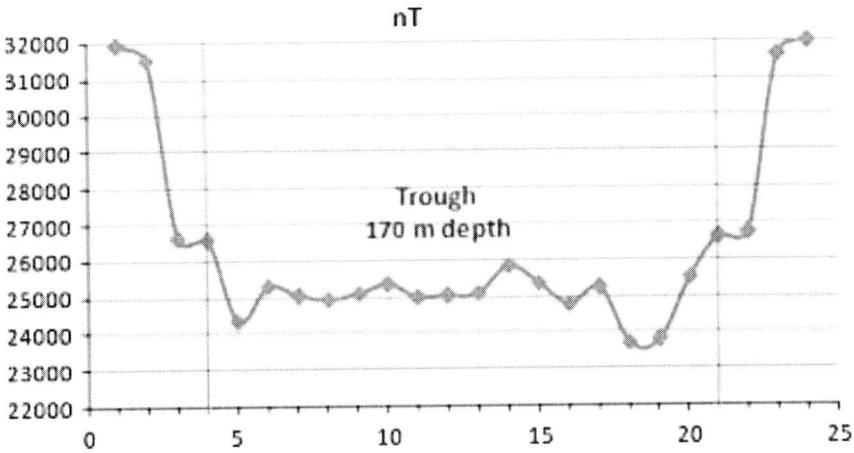

Figure 9. A magnetic profile showing the graben-like structure along the LST anomalies in El-Gallaba Plain [44].

Consequently, these structurally controlled linear basins could be promising areas for groundwater accumulation and exploration in the El-Gallaba Plain of the Western Desert in Egypt [44].

Nile Delta and Nile Valley

The hydrogeological mapping at a scale of 1:100,000 covers the densely populated areas of Egypt, including the Nile Valley, Nile Delta and their desert fringes, the New Valley, the desert Oases, and portions of the coastal zone. The hydrogeological map of the Nile Delta was completed in the period 1989–1992 produced by RIGW within the framework of the main project entitled: "Development and management of groundwater resources in the Nile Valley and Delta". This project was an Egyptian – Dutch bilateral project. Examples of these maps include RIGW/IWACO [20]:

–For Nile Valley: Sohag – Assiut – Girga – Luxor – Esna – Sohag (1990) –....

–For Nile Delta and desert fringes: Tanta – Mennouf – Zagazig – Ismailia – Wadi El-Natrun (1990) – ...

These detailed maps are used in different applications; rising water table, influence of groundwater table on monuments, assessment of groundwater vulnerability to pollution (RIGW/IWACO) [20].

Dawoud et al. [45] developed a GIS-based model to simulate the water resources in the Western Nile Delta. This model has provided a useful tool to evaluate the water resources status and to test the proposed management alternatives in the Western Nile Delta region. The model consisted of multi-layers aquifer system. GIS was used to manage the spatially distributed input parameters such as the time invariant spatial data and outputs of the model. The model was calibrated against the available average annual groundwater heads of about 60 observed wells during the period from 1990 to 2002. The calibrated model has been used to test two management alternatives.

CONCLUDING REMARKS AND RECOMMENDATIONS

From the previously mentioned studies related to the use of RS and GIS in groundwater mapping, it could be concluded that groundwater mapping is one of the main tools for efficient and controlled development of groundwater resources. These maps are used by engineers, planners and decision makers to allocate, develop and manage groundwater within a national water policy. Integrated RS and GIS techniques prove to be rather satisfactory options for groundwater mapping in different regions of Egypt.

From these studies, the following concluding remarks could be highlighted:

a) Integrated RS data and GIS are useful in delineating favorable areas for well establishment using various geo-informative thematic maps.
b) Modeling approach is applied to integrate the physical and geologic factors governing groundwater potentiality in a map form showing spatial data for areas with different responses to groundwater potentiality in the Sinai Peninsula.
c) Interpretation of Radarsat images and high-resolution SIR-C radar scenes combined with DEM analysis shows that NE and ENE drainage directions prevail in south-western Egypt and culminate in the Kharga Depression. These results will be helpful in strategic planning for ground-water exploration and drilling programs.

d) Information has been extracted on lithology, geological structures, landforms, land use/land cover, soil characteristics, slope of the terrain and studying all these in an integrated way in a GIS environment in order to generate village-wise groundwater prospect zones.

As a future approach, within the field of groundwater mapping in Egypt, it is highly recommended to:

a) Update the detailed hydrogeological maps of Egypt at a scale of 1:100,000, produced by RIGW for the Nile Delta and Nile Valley using RS and GIS techniques. Maps updating is recommended for assessing the actual short and long-term changes of groundwater regime caused by over pumping and population growth needs in these areas.
b) Produce detailed groundwater maps of the Western Desert Oasis for the purpose of supporting future development plans. Moreover, Radar images may provide significant advantages for groundwater mapping in arid areas.
c) Take the approach of launching an Egyptian satellite specified in mapping Desert areas within an interdisciplinary collaboration between Egypt and the neighboring Arab and African countries.
d) Much more RS and GIS based groundwater research with field investigations may be tried to effectively exploit the current and expanding potential of the technologies.

ACKNOWLEDGMENTS

The author is very grateful to the anonymous referees for their critical evaluation of the manuscript and their instructive comments. The author would also like to thank Prof. El-Sayed Zaghloul and Prof. Abdalla Gad for providing constructive comments and suggestions which improved the manuscript.

REFERENCES

1. United Nations, Department of Economic and Social Affairs, Population Division. World population prospects: the 2010 revision, highlights and advance tables. ESA/P/WP.220; 2011.

2. Idris H, Nour S. Present groundwater status in Egypt and the environmental impacts. Environ Geol Water Sci 1990;16(3):171–7.
3. Food and Agriculture Organization. Review of world water resources by country. Rome, Italy: Food and Agriculture Organization of the United Nations; 2003.
4. Bear J. Hydraulics of groundwater. New York: McGraw-Hill; 1979.
5. Jha MK, Chowdary VM. Challenges of using remote sensing and GIS in developing nations. Hydrogeol J 2007;15:197–200.
6. El Tahlawi MR, Farrag AA, Ahmed SS. Groundwater of Egypt: "an environmental overview". Environ Geol 2008;55:639–52.
7. UNESCO. Joint programme for climate change risk management in Egypt – proposed climate change adaptation strategy for the Ministry of Water Resources & Irrigation Egypt. Ministry of Water Resources and Irrigation; 2013.
8. MWRI. Water resources development and management strategy in Egypt – 2050. Cairo, Egypt: Ministry of Water Resources and Irrigation; 2010 in Arabic..
9. El Arabi N. Environmental management of groundwater in Egypt via artificial recharge extending the practice to Soil Aquifer Treatment (SAT). Int J Environ Sustain 2012;1(3):66–82, ISSN 1927-9566.
10. Allam AR, Saaf Ele-Jan, Dawoud MA. Desalination of brackish groundwater in Egypt. Desalination 2002(152):19–26.
11. Alnaggar D. Water resources management and policies for Egypt. Workshop on policies and strategies options for water management in Islamic countries, Tehran; December 2003.
12. RIGW/IWACO. Hydrogeological mapping of Egypt, scale 1:2,000,000. 1st ed. TN 70.120-88-03; 1988.
13. RIGW/IWACO. Contributions to environmental management of Egypt's groundwater resources. Final report, EMGR Project, Cairo; 1999.
14. Dawoud MA. Design of national groundwater quality monitoring network in Egypt. Environ Monit Assess 2004;96:99–1180.
15. Hefny K, Attia F. Groundwater in the present and future. In: Proceedings of the national conference on the future of land reclamation in Egypt, Cairo; 1990.
16. Attia FAR. The groundwater plan, national level. Giza, Egypt: MWRI; 2001.
17. Shata AAO. The groundwater resources in the coastal zone. Preliminary report, Part 1. Cairo: Water resources Department, Desert Institute; 1978.
18. Khalil J, Mebius J. The role of hydrogeological mapping in groundwater development. In: Water science no. 10, Special issue on the Round Table meeting on planning for groundwater development in arid regions. Cairo: WRC; 1991.

19. UNESCO. International legend for hydrogeological maps, revised ed.; 1983.
20. RIGW/IWACO. Assessment of groundwater vulnerability. TN 70.130-89-02; 1989.
21. UNESCO. Capacity building of The Egyptian Geological and Mining Authority (EGSMA) and The National Authority for Remote Sensing and Space Sciences (NARSS) for the sustainable development of the South valley and Sinai in cooperation with UNDP and UNESCO EGY/97/011; 2007.
22. Becker MW. Potential for satellite remote sensing of ground water. Ground Water 2006;44(2):306–18.
23. Meijerink AMJ. Remote sensing applications to groundwater. IHP-VI, series on groundwater no. 16, UNESCO 2007; 2007.
24. Jasmin I, Mallikarjuna P. Review: satellite-based remote sensing and geographic information systems and their application in the assessment of groundwater potential, with particular reference to India. Hydrogeol J 2011;19:729–40.
25. Lillesand TM, Kiefer RW. Remote sensing and image interpretation. Toronto: John Wiley and Sons; 2000.
26. Lillesand TM, Kiefer RW, Chipman JW. Remote sensing and image interpretation. 5th ed. John Wiley and Sons; 2004.
27. Zevenbergen AW, Radstake F, Tusveld M. Remote sensing and GIS applications for hydrogeological studies in Egypt. Planning for groundwater development in arid and semi-arid regions, Round Table Meeting (RTM-91), Cairo; 5–9 October 1991.
28. Kwarteng AY. Utilization of remote sensing and GIS for groundwater exploration in Kuwait. In: Sherif, Singh, Al-Rashed, editors. Groundwater hydrology. The Netherlands: Balkema; 2002. p. 157–78.
29. Jaiswal RK, Mukherjee S, Krishnamurthy J, Saxena R. Role of remote sensing and GIS techniques for generation of groundwater prospect zones towards rural development – an approach. Int J Remote Sensing 2003;24(5):993–1008.
30. Oh Hyun-Joo, Kim Yong-Sung, Choi Jong-Kuk, Park Eungyu, Lee Saro. GIS mapping of regional probabilistic groundwater potential in the area of Pohang City, Korea. J Hydrol 2011;399:158–72.
31. Droogers Peter, Immerzeel Walter, Perry Chris. Application of remote sensing in national water plans: demonstration cases for Egypt, Saudi-Arabia and Tunisia. Future Water 2009(February).
32. Robinson CA, El-Baz F, Al-Saud TSM, Jeon SB. Use of radar data to delineate palaeodrainage leading to the Kufra Oasis in the eastern Sahara. J Afr Earth Sci 2006;44(2006):229–40.

33. Omran AA. Integration of remote sensing, geophysics and GIS to evaluate groundwater potentiality – a case study in Sohag Region, Egypt. In: The 3rd international conference on water resources and arid environments and the 1st Arab Water Forum; 2008.
34. Abdalla F. Mapping of groundwater prospective zones using remote sensing and GIS techniques: a case study from the Central Eastern Desert, Egypt. J Afr Earth Sci 2012;70(2012):8–17.
35. Gheith H, Sultan M. Construction of a hydrologic model for estimating Wadi runoff and groundwater recharge in the Eastern Desert, Egypt. J Hydrol 2002;263:36–55.
36. Amer R, Ripperdan R, Wang T, Encarnacio′n J. Groundwater quality and management in arid and semi-arid regions: case study, Central Eastern Desert of Egypt. J Afr Earth Sci 2012;69:13–25.
37. Japan International Cooperation Agency (JICA). North Sinai groundwater resources study in the Arab Republic of Egypt. Final report submitted to the Research Institute for Water Resources. Ministry of Public Works and Water Resources, Egypt; 1992.
38. Japan International Cooperation Agency (JICA). South Sinai groundwater resources study in the Arab Republic of Egypt, Tokyo, Japan; 1999.
39. Elewa HH, Qaddah AA. Groundwater potentiality mapping in the Sinai Peninsula, Egypt, using remote sensing and GISwatershed-based modeling. Hydrogeol J 2011;19:613–28.
40. Sultan A, Abdel Rahman N, Ramadan TM, Salem SM. The use of geophysical and remote sensing data analysis in the groundwater assessment of El Qaa Plain, South Sinai, Egypt. Aust J Basic Appl Sci 2013;7(1):394–400.
41. Milewski A, Sultan M, Jayaprakash SM, Balekai R, Becker R. RESDEM, a tool for integrating temporal remote sensing data for use in hydrogeologic investigations. Comput Geosci 2009;35:2001–10.
42. Robinson C. Application of satellite radar data suggest that the Kharga Depression in south-western Egypt is a fracture rock aquifer. Int J Remote Sensing 2002;23(19):4101–13.
43. Robinson C. Understanding the distribution of groundwater resources using Synthetic Aperture Radar Data over Southwest Egypt. In: Geoscience and remote sensing symposium, 2008. IGARSS 2008. IEEE International. vol. 3, Boston, MA; 7–11 July 2008.
44. Koch M, Gaber A, Gereish MH, Zaghloul E, Arafat SM, AbuBakr M. Multisensor characterization of subsurface structures in a desert plain area in Egypt with implications for groundwater exploration. In: SPIE remote sensing conference, Dresden, Germany; 23–26 September 2013.

45. Dawoud MA, Darwish MM, EL-Kady MM. GIS-based groundwater management model for Western Nile Delta. Water Resour Manage 2005;19:585–604.

CITATION

Salwa Farouk Elbeih, An overview of integrated remote sensing and GIS for groundwater mapping in Egypt, Ain Shams Engineering Journal, Volume 6, Issue 1, March 2015, Pages 1-15, ISSN 2090-4479, http://dx.doi.org/10.1016/j.asej.2014.08.008.

CHAPTER 7

Satellite Remote Sensing Outputs of the Certain Glaciers on the Territory of East Georgia

G. Kordzakhia[1], L. Shengelia[1], G. Tvauri[2], V. Tsomaia[1], M. Dzadzamia[3]

[1] Georgian Technical University, The Institute of Hydrometeorology, Georgia

[2] I. Javakhishvili Tbilisi State University, M. Nodia Institute of Geophysics, Georgia

[3] National Environmental Agency of the Ministry of Environment and Natural Resources Protection of Georgia, Georgia

ABSTRACT

The variations in glaciers are the important indicators of regional climate change. The glaciers play an important role in the regulation of water balance. In the conditions of global warming they recede and degrade that is expressed in the related changes in glacier runoff. The research of glacier melting is important for studies of sea/ocean level changes. The Caucasian glacial dimensions (area, volume, length) have been changed over the centuries.

The scientific study of glaciers in the Caucasus was started during the first half of 18th century. In the last century the terrestrial observations on glaciers were carried out. Due to the difficulties of organization and conducting of field works the received observational data sets consists from different series of various temporal duration characteristics of glaciers. The data received contain uncertainties. From another hand carrying out of such field works are expensive.

With the launch of the Earth's satellites it was determined that satellite remote sensing is the best technology allowing to receive data with needed regularity in terms of both time and space resolution. Some uncertainties remain in the data as the observational tool is too far away from the Earth's surface. So, the necessity for the strong quality assessment/quality control (QA/QC) remains. A lot of studies showed that the best method for investigation of glaciers is application of satellite remote sensing combined with terrestrial observations and expert knowledge of separate glaciers.

INTRODUCTION

It is obvious that variations of glaciers are clear indicators of regional climate change. Recent interest in the study of glaciers is motivated by their clear reaction to climate changes. As glaciers cover so many climate regions worldwide, from the tropics to poles, their study enables a survey of significant aspects of global and regional Climate Change. As many of their features are detectable from the space, such investigations are well adapted for the methods of remote sensing. The glaciers play an important role in the regulation of water balance in certain regions. Under the global warming glaciers recede and degrade, which is reflected in the related changes of glacier runoff. These processes result on glacial and hydrological disasters such as ice blocks falls from the glaciers accompanied with the debris, followed by the river bed blockage and natural dam formation, with consequent break and pass of catastrophic flash floods and/or mudflows (Kääb et al., 2006).

The investigation of glacier melting is important for studies of sea/ocean level changes that also may have a significant risk for the residents of coastal areas. That is one of the most negative impacts of modern climate change on mankind. The abovementioned put into agenda the necessity for detailed study of the glaciers.

The scientific study of glaciers in the Caucasus was started during the first half of 18th century. In the last century the terrestrial observations on glaciers were carried out. Due to the difficulties of organization and conducting the field works the terrestrial observation data resulted in different series of various temporal duration characteristics of glaciers. The data received contain uncertainties. From another hand carrying out of field works are expensive and data gathered have no sufficient spatial and temporal resolution.

With launch of the Earth's satellites it was determined that satellite remote sensing is the best technology allowing to receive data with needed regularity in terms of both time and space resolution. Some uncertainties remain in the data as the observational tool is too far away from the Earth's surface. So, the necessity for the strong quality assessment/quality control (QA/QC) remains. A lot of studies showed that the best method for investigation of glaciers is application of satellite remote sensing combined with terrestrial observations and expert knowledge of separate glaciers.

RESEARCH AREA & GLACIERS SURVEY

Georgia is a transcontinental country, located in the South Caucasus, between the Black Sea to the West and the Caucasus mountains to the North. Georgia is bordered by Russian Federation from the North; it is sided from the South by Turkey and Armenia and it is bordered from the South-east by Azerbaijan. The territory of the country is divided by Likhi Range into two parts: West Georgia and East Georgia. This division is corroborated by the difference in physical and geographical conditions between them as well. The glaciers are located in North part of the country in the Caucasus Mountains. The main glaciers from the Caucasus Mountains in the territory of East Georgia are researched in the present study. Below are listed the important glaciers to be studied based on the satellite remote sensing. The main objective of the research is the adjustment of characteristics of Mayly, Chachy, Devdoraki, Abano, Gergety, Mna, East Suatisi, Middle Suatisi, West Suatisi, SU4G08011072, SU4G08011061, SU4G08011056, SU4G08011059 and SU4G08011058 glaciers. The last 5 glaciers have no names and they are registered in World data base as mentioned above i.e. SU4G08011072, SU4G08011061, SU4G08011056, SU4G08011059 and SU4G08011058.

In Fig. 1 as an example Aster DEM and Landsat satellite images are presented for Kazbegi glacial mountain array.

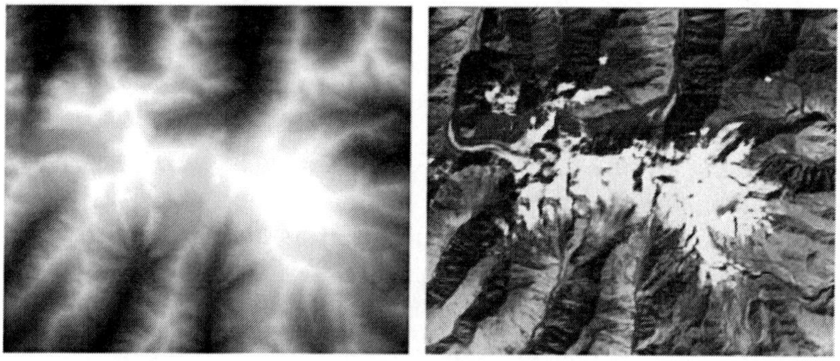

Figure 1. Aster DEM and false colour Landsat images.

In Fig. 2 the 3D visualization of Landsat images and Aster DEM (left) and Google Earth image (right) of Kazbegi glacial mountain array are presented.

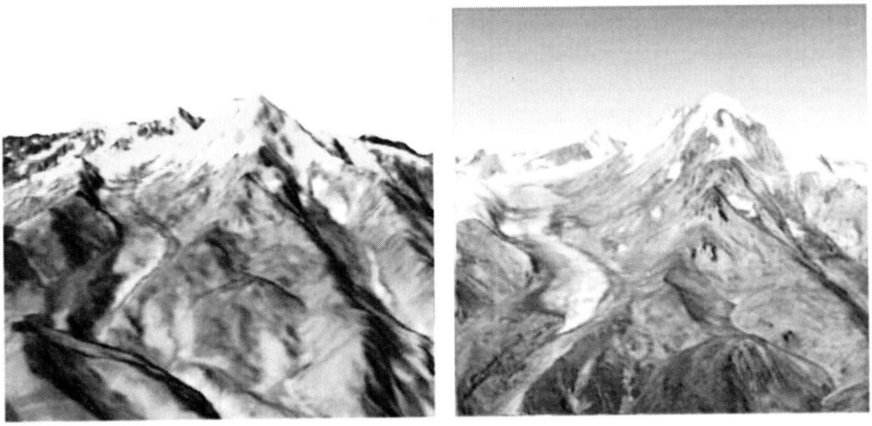

Figure 2. Kazbegi Glacier's Array 3D Visualization of Landsat images and Aster DEM (left) and Google Earth image (right).

The satellite remote sensing gives possibility of universal, systematic and complex research of glaciers properties and changes. This is due to the fact that satellite remote sensing allows to cover efficiently the interesting regions and to study dynamically such important parameters, as the glacier area, terminus positions at the end of its location, firn line elevation, accumulation area, the hypsometry (Pellikka and Gareth, 2010). In turn these parameters allow calculation of glacier mass balance.

In Georgia the glacial parameters (area, volume, length) have been changed over the centuries. It is determined that from the second half of the last century the characteristics of the Georgian glaciers are steadily diminishing. The process is still underway and will likely continue in the future. The elevation of the lower boundary of the glaciers increases.

During the last century the area of the Georgian glaciers decreased by 36% and the volume reduced by 48%. Some glaciers melted away but the glacier melting process causes increase of glacier number due the fragmentation of large glaciers into small ones. Hence, the total number of glaciers increased. The length of glaciers was reduced by 600 m in average. In the lower part of glaciers ice thickness decreased by 50–150 m and in the upper part it was reduced by 20–30 m consequently. During the last decades (1970–2000) glacial conditions in Georgia have changed on the background of rainfall and temperature increase by 10–15% and 1 °C respectively. During the same period the glacier area decreased by 12%, the ice volume by 15%, while the number of glaciers increased by 2.4% (Stokes et al., 2006).

Recent studies proved sustainable recession of the Georgian glaciers. For example, in the summer of 2010 the Gergeti glacier tongue amounted up to 8.5 km. From 2004 up to 2010 it decreased by 123 m in average of about 20 m per year (Keggenhoff et al., 2011). During the period of 1810–2011 the melting dynamics of the glacier Tviberi is as follows: in 1810 its area was 47.5 km^2, and in 1889 – 47.1 km^2. Simple analysis shows that glacier area during 79 years decreased by 1%. In 1965 its area amounted to 41.6 km^2, i.e. the glacier area decreased by 12% compared to 1889 and in 2011 it was reduced by 35% compared to 1889 and was equal to 30.9 km^2. A similar case occurs with glacier Chalaati. In 1965 its area was 14.4 km^2 and in 2011 – 11.8 km^2, i.e. the glacier area decreased by 16% for the last 46 years (Shengelia et al., 2012).

RESEARCH METHODS

The following actions are performed for specifying the contours of the glaciers according to the developed methodology:

- The Google software products are used. Namely, the Google Earth satellite images and the satellite sensor ASTER data based on the generated Digital Terrain Model (Aster DEM) are applied. GIS systems: Google Earth, BEAM Visat and Quantum GIS Lisboa are used for satellite data processing. It would be noted that the spatial resolution of satellite Aster DEM is 30 m (Bolch and Kamp, 2005);
- Various glacier characteristics stored in the GLIMS database: glaciers contours and numerical characteristics of glaciers are used for the glacier identification. GLIMS-database protected contours are generated based on the Aster sensor data (Raup et al., 2007);
- Soviet topographic maps (1: 50,000) as well as the terrestrial observations data are used for glacier contour validation. It is clear that such data exist only for the certain glaciers and for some years;
- Expert knowledge is also effectively used for the investigation of these glaciers.

Changes in glacier area and terminus positions have been widely used as the indicator of a glacier response to climate impact. These two parameters are relatively easy to determine from multispectral satellite images. Glacier outlines combined with a digital elevation model (DEM) give possibility to determine such glacier parameters as hypsometry, minimum, maximum and mean elevations, firn line altitude.

Delineation of glacier outlines is possible both manually and automatically based on the corresponding software by the band combination techniques. Human interpretation remains the best tool for extracting high level information from satellite imagery. Tedious, manual digitalization of glacier boundaries by an operator with good knowledge of the region can produce glacier boundary contours of high quality and accuracy. Automatic delineation of glacier relies on different spectral properties of glacial ice and snow in the visible and IR part of electromagnetic spectrum. The greatest difficulty in glacier mapping from remote sensing is the presence of debris cover on glaciers. In such cases higher resolution satellite images and DEM are used as well as the data of the terrestrial observations and expert knowledge of separate glaciers.

RESULTS

For 3D visualization of Kazbegi Glacier's array Landsat TM image (path: 170, row: 30, level L1T) from 26 September of 1984 year and ASTER DEM edition images are used. Landsat data were downloaded from USGS Earth explorer webpage, consequently ASTER DEM – from ASTER webpage of NASA JPL website. Both of them may be downloaded free of charge, in GeoTIFF format and used in any GIS 3D visualization of glaciers from study area. Kazbegi glaciers as well as all researched glaciers on the territory of East Georgia are performed by generating the Landsat false colour composite images (band 5 as red, 4 as green and 3 as blue) and wrapping over the ASTER DEM.

The boundaries of glacier array were adjusted using the high spatial resolution satellite images, e.g. for the Kazbegi glaciers based on the satellite image dated September 1, 2010, provided by Google Earth software package. That allows high accuracy digitalization of the contours of all the glaciers. Digitalization of glacier contours is performed manually on the basis of expert knowledge. Corresponding shapefiles are created. Results are compared with GLIMS database contours. These contours are derived from processing of Aster satellite images (e.g. for the Kazbegi glaciation based on the satellite image dated September 25, 2004, DEM spatial resolution – 15 m) by scientists of the Institute of Geography of Russian Academy of Sciences.

Further, Gergeti glacier will be considered as an example for the simplicity.

In Fig. 3 contours of Gergeti glacier are presented. Blue contour corresponds to the Gergeti glacier contour from GLIMS database. By red one is shown Gergeti glacier adjusted contour using satellite remote sensing. As it is revealed in Fig. 3 the difference between these contours is significant, especially in the Northern part. The reason of this difference is detected based on DEM analysis. In GLIMS data base (see Fig. 3) the whole field of the northern slope of the Kazbegi mountain is assigned to the Gergeti glacier. Manual digitalization of Gergeti glacier contour, is based on Digital Globe high resolution satellite image, as well as applying ASTER DEM data. Using the Quantum GIS software package elevation isolines with 30, 10 and 5 m resolution are generated based on ASTER DEM. The outputs are plotted, as additional layer to the Google Earth. As a result glacier contours are adjusted taking into account the watershed on the North-west slope of Gergeti glacier. The analysis shows that the platue part from the adjusted red line till the blue line from the GLIMS data base should be attributed to Devdoraki glacier.

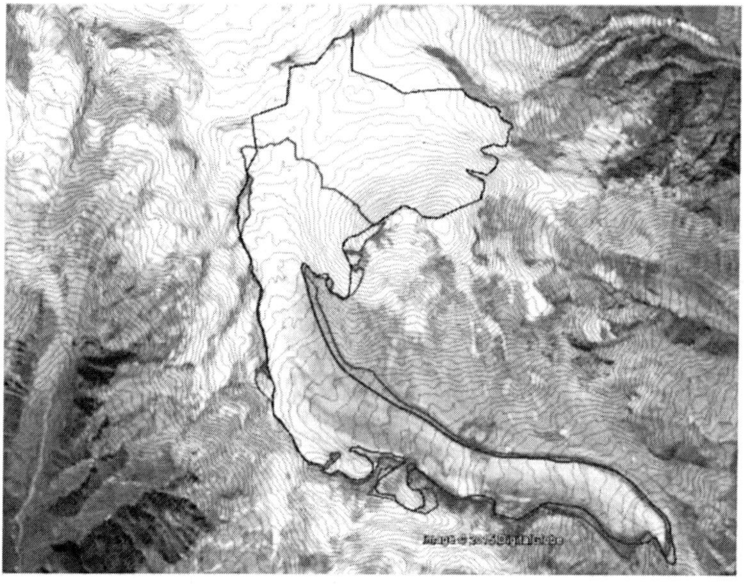

Figure 3. Gergeti glacier with initial (blue line) and adjusted (red line) contours. Isoline elevation step is equal to 30 m.

Fig. 4 illustrates distribution of DEM pixel numbers for Gergeti glacier according to the elevation. For determination of hypsometric curve adjusted outlines of Kazbegi glaciers and ASTER DEM data BEAM-VISAT are used.

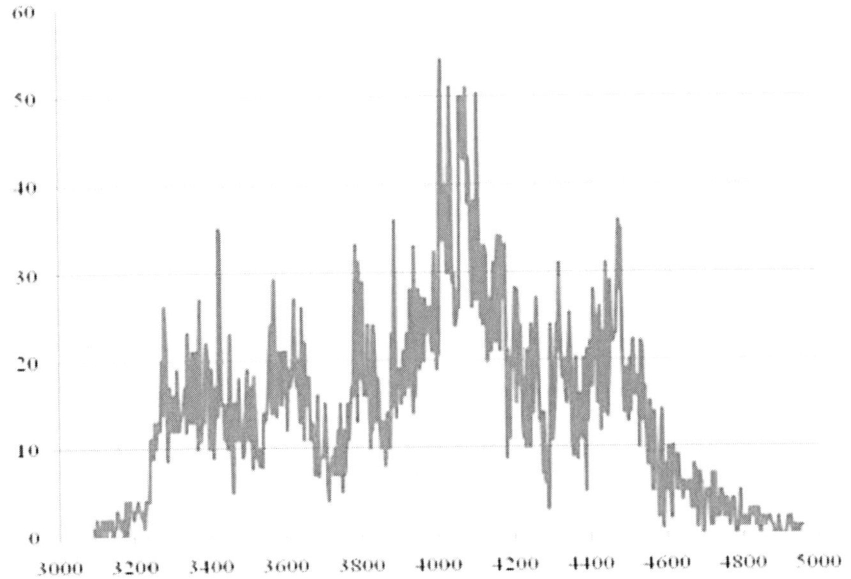

Figure 4. Distribution of the DEM Pixel Numbers for Gergeti Glacier according to the elevation.

The firn line elevation was determined according to the Hefer method (Tsomaia and Drobishev, 1977) and ablation area was calculated through the hypsometry curve (Racoviteanu et al., 2008).

According to the Hefer method the elevation of the firn line is the average of the firn line basin height and the elevation of the glacier tongue. The average of the values of the highest peak elevation (two or more peaks) around the glacier basin is determined and the value of the elevation of glacier tongue is added. Finally value of the firn line elevation is the average of these two values.

Glacier accumulation area can be determined using hypsometric curve. Glacier accumulation area is the value of one pixel area multiplied by the total number of pixels on hypsometric curve from the firn line elevation till the highest height of the glacier. Glacier ablation area is the value of one pixel area multiplied by the total number of pixels on hypsometric curve from the firn line elevation till the lowest height of the glacier. The sum of the glacier accumulation and ablation areas makes a total glacier area.

In Fig. 5 the 3D visualization of Gergeti glacier is shown. Blue contour corresponds to the Gergeti glacier contour from GLIMS database. By red one is shown Gergeti glacier adjusted contour using satellite remote sensing. The green line shows the areas free of debris. It is natural that green line differs from the blue line in the lower part of the glacier as the upper part is free of debris.

Figure 5. The 3D visualization of Gergeti glacier. The blue contour corresponds to the Gergeti glacier outline from GLIMS database, by red line is shown Gergeti glacier outline adjustment using satellite remote sensing data. The green line shows the areas free of debris.

The comparison of Gergeti glacier contours received from the satellite remote sensing with Soviet topographic map (1:50,000) is presented in Fig 6. Comparison of topographic map with contours (Fig. 6) shows the changes of contour of Gergeti glacier.

These data are compared with the glacier relevant information from the USSR glaciers catalogue (Tsomaia and Drobishev, 1977).

In Table 1 the glacier parameters, such as length, area, minimum elevation, firn line elevation, ablation area of the listed glaciers of East Georgia calculated from the satellite remote sensing data (columns marked by (1))

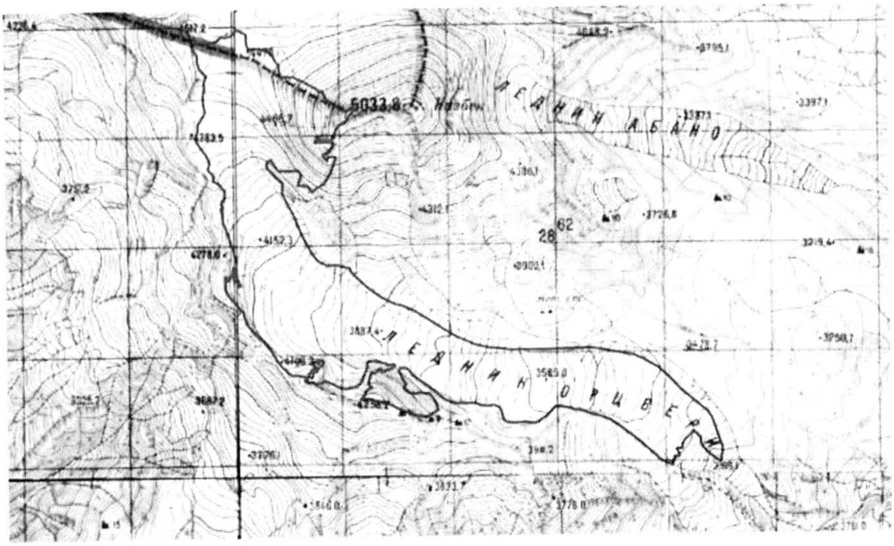

Figure 6. The comparison of Gergeti Glacier contours received from the satellite remote sensing with topographic map.

and from the USSR glaciers catalogue (columns marked by (2)) are presented. This table allows comparison of the data received from satellite remote sensing with the data from USSR glaciers catalogue.

Table 1. Characteristics of certain glaciers located on the territory of East Georgia. Columns marked by (1) present the data received from the satellite remote sensing and columns marked by (2) show the data from USSR glacier catalogue.

No.	Glacier name	WGI Glacier ID	Local ID	Maximum length (km)				Area (km^2)				Minimum elevation, (m)				Maximum elevation, (m)		Firn line elevation (m)		Ablation area (km^2)			
				Whole glacier		Free of debris		Whole glacier		Free of debris		Whole glacier		Free of debris						Whole glacier		Free of debris	
				(1)	(2)	(1)	(2)	(1)	(2)	(1)	(2)	(1)	(2)	(1)	(2)	(1)	(2)	(1)	(2)	(1)	(2)	(1)	(2)
1	Mayly	SU4G08010041	248	5.83	6.2	5.64	6.2	6.43	7.0	6.06	7.0	2,469	2,360	2,521	2,360	4,610	4,600	3,225	3,170	1.02	1.4	0.69	1.4
2	Chachy	SU4G08011046	242	3.22	3.2	3.1	3.2	2.28	2.8	2.07	2.8	3,239	3,230	3,299	3,230	4,447	4,430	3,665	3,660	0.78	0.6	0.53	0.6
3	Devdoraki	SU4G08011048	241	6.87	7.3	6.321	6.5	6.24	7.0	6.04	6.6	2,392	2,260	3,549	2,540	5,038	5,030	3,326	3,260	0.74	1.0	0.48	0.6
4	Abano	SU4G08011049	240	3.83	4.1	2.26	2.5	1.68	2.0	1.00	1.4	3,020	2,950	3,462	3,410	5,038	5,030	3,735	3,700	0.87	0.9	0.18	0.3
5	Gergety	SU4G08011052	238	7.1	8.5	7.1	8.5	5.15	8.3	4.63	7.1	3,091	2,870	3,091	2,870	4,936	5,030	3,793	3,650	1.50	2.3	1.20	1.1
6	None	SU4G08011056	235		1.9	0.973	1.4		0.7	0.37	0.6		3,350	3,758	3,460	4,227	4,220	3,884	3,680		0.3	0.06	0.2
7	Denkera	SU4G08011057	234		2.3	1.324	2.3		2.4	0.42	2.0		3,500	3,655	3,500	4,103	4,230	3,847	3,770		0.8	0.23	0.4
8	None	SU4G08011058	234a		1.0	1.808	1.0		0.2	0.52	0.2		3,570	3,567	3,570	4,285	4,200	3,818	3,820		0.1	0.13	0.1
9	None	SU4G08011059	233a		2.3	1.775	2.3		0.7	0.75	0.7		3,440	3,450	3,440	4,327	4,330	3,765	3,760		0.3	0.23	0.3
10	Mna	SU4G08011060	233	3.32	4.1	2.43	3.8	3.136	4.6	2.95	4.3	3,036	2,860	3,330	3,060	4,612	4,600	3,568	3,480	0.36	0.6	0.17	0.3
11	None	SU4G08011061	232	4.337	3.2	2.063	3.2	2.367	2.1	1.21	2.1	3,292	3,330	3,378	3,330	4,586	4,310	3,681	3,700	1.19	0.8	0.16	0.8
12	East Suatisi	SU4G08011062	231	4.45	5.4	4.45	5.4	7.927	10.2	7.45	10.0	3,230	3,000	3,230	3,000	4,532	4,580	3,615	3,500	3.019	3.3	2.23	3.1
13	Middle Suatisi	SU4G08011063	230	3.98	4.7	3.62	4.7	2.123	2.5	2.05	2.5	2,975	2,850	3,082	2,850	4,734	4,760	3,583	3,520	0.560	0.6	0.52	0.6
14	West Suatisi	SU4G08011064	229	2.95	3.5	2.95	3.5	1.832	2.4	1.779	2.4	3,284	3,070	3,284	3,070	4,444	4,460	3,707	3,600	0.502	0.4	0.46	0.4
15	None	SU4G08011072	222	1.249	1.4	1.249	1.4	0.427	0.6	0.427	0.6	3,334	3,150	3,334	3,150	3,909	3,920	3,582	3,490	0.14	0.2	0.14	0.2

CONCLUSIONS

The glacier research is significant due to the fact that the glaciers occupy approximately 11% (16 million square kilometres) of the earth's land area. The glaciers contain a large number of fresh-water (30 million km^3), which is the 2/3 of the earth's fresh water supply. Georgia possesses more than 600 glaciers and their detailed study is important from several points of view. In present research satellite remote sensing research is done for the main glaciers located on the territory of East Georgia.

The scientific study of glaciers in the Caucasus was started during the first half of 18th century. In the last century the terrestrial observations on glaciers were carried out. Due to the difficulties of organization and conducting the field works the terrestrial observations data resulted in different series of various temporal duration characteristics of glaciers. The data received contain an uncertainties. From another hand carrying out of field works are expensive and data gathered have no sufficient spatial and temporal resolution.

With launch of the Earth's satellites it was determined that satellite remote sensing is the best technology allowing to receive data with needed regularity in terms of both time and space resolution. Some uncertainties remain in the data as the observational tool is too far away from the Earth's surface. So, the necessity for the strong quality assessment/quality control (QA/QC) remains. A lot of studies showed that the best method for investigation of glaciers is application of satellite remote sensing combined with terrestrial observations and expert knowledge of separate glaciers.

In cases whenever the terrestrial observations are not available and expert knowledge is absent, validation is not possible. In our research maximum length, area, minimum elevation are not determined for some glaciers, namely: SU4G08011056, SU4G08011057, SU4G08011058 and SU4G08011059. Accordingly only some information about those glaciers is presented but the detailed conclusions about them is not possible to make at that stage.

Based on the detailed comparison of the output data received using the satellite remote sensing provided with the QA/QC with the similar data from the glaciers catalogue it can be concluded that during the last

50 years all the characteristics of the main glaciers from the East Georgia are changing as follows:

- The minimum elevation and equilibrium line altitude are increasing;
- The length and area as well as the ablation area are decreasing.

The carried researches define that impact of modern climate change is well expressed in East Georgian glaciers retreat due to modern warming.

It would be noted that the complex research of the glaciers using different sources' data such as satellite remote sensing, terrestrial observations and expert knowledge is effective for better accurate determination of glaciers' main characteristics.

ACKNOWLEDGMENT

The research is carried in frames of Shota Rustaveli National Foundation of Georgia that supported research grant project FR/586/9-110/13.

REFERENCES

1. Bolch, T., Kamp, U., 2005. Glacier mapping in high mountains using DEMs, Landsat and ASTER data, Grazer Schriften der Geographie und Raumforschung. In: Proceedings 8th International Symposium on High Mountain Remote Sensing Cartography, vol. 41, 20–27 March 2005. La Paz, Bolivia, pp. 13–24.
2. Ka¨a¨b, A., Huggel, C., Fischer, L., 2006. Remote sensing technologies for monitoring climate change impacts on glacier- and permafrostrelated hazards. In: ECI Conference on Geohazards, Lillehammer, Norway. pp. 1–10.
3. Keggenhoff, V.I., Kellerm, T., Elizbarashvili, M., Gobejishvili, R., King, L., 2011. Naturkatastrophen durch Klimawandel im Kaukasus? Hochwasser und Hanginstabilitaten in Georgien immer haufiger. Spiegel der Forschung N2, 16–23.
4. Pellikka, Petri, Gareth Rees, W., 2010. Remote Sensing of Glaciers. Taylor & Francis Group, London, UK. Racoviteanu, A.E., Williams, M.W., Barry,

R.G., 2008. Optical remote sensing of glacier characteristics: a review with focus on the Himalaya. Sensors 8 (5), 3355–3383.

5. Raup, B.H., Racoviteanu, A., Khalsa, S.J.S., Helm, C., Armstrong, R., Arnaud, Y., 2007. The GLIMS geospatial glacier database: a new tool for studying glacier change. Global Planet. Change 56, 101–110.
6. Shengelia, L., Kordzakhia, G., Tvauri, G., Davitashvili, T., Begalishvili, N., 2012. Possibilities of the use of remote sensing technologies for the estimation of modern climate change impact on the Caucasus glaciers. Monthly Scientific-Reviewed Magazine of Georgian National Academy of Sciences, Metsniereba da Technologiebi (Science and Technologies), No. 4–6, pp. 25–30 (in Georgian). Stokes, C.R., Gurney, S.D., Shahgedanova, M., Popovnin, V., 2006.
7. Late-20th-century changes in glacier extent in the Caucasus Mountains. J. Glaciol. 176, 99–109.
8. Tsomaia, V.S., Drobishev, O.A., 1977. Glaciers Catalog of the USSR, the North Caucasus, Leningrad, vol. 8. Gidrometeoizdat, p. 71 (In Russian).

CITATION

G. Kordzakhia, L. Shengelia, G. Tvauri, V. Tsomaia, M. Dzadzamia, Satellite remote sensing outputs of the certain glaciers on the territory of East Georgia, The Egyptian Journal of Remote Sensing and Space Science, Available online 29 June 2015, ISSN 1110-9823, http://dx.doi.org/10.1016/j.ejrs.2015.06.003.

Index

A

Airworthiness, 13, 14, 15
Algal bloom proxie, 69, 72
American Society for Photogrammetry and Remote Sensing (ASPRS), 5, 55
artificial intelligence, 143

B

Biliproteins, 79

C

Chlorophyll concentrations, 74, 78, 100, 105
crop requirement, 119, 121, 136, 137, 142

D

Danish Transport Authority, 16
Digital Terrain Model (DTM), 119, 125, 130
Discrimination, 70, 84, 106, 108

E

Earth's satellite, 233, 234, 243
Earth's surface, 233, 234, 243
electromagnetic measurement, 168, 174
Empirical blue-green, 72
erosion risk, 119, 127, 139

F

freshwater resource, 201

G

genetic programming, 144, 146, 159, 160, 161, 162, 163, 164
Geomatic community, 9
glaciers cover, 234
Global Navigation Satellite Systems (GNSS), 9
Ground control station (GCS), 2
Ground Control Stations (GCSs), 10
groundwater flow, 167, 168, 170, 171, 173, 174, 178, 185, 188, 191, 192, 193, 194, 195, 198, 199
groundwater management, 201, 202, 232
groundwater mapping, 201, 202, 206, 210, 227, 228, 232

groundwater resources, 201, 206, 207, 214, 217, 224, 226, 227, 229, 231

H

horizontal agricultural expansion, 119, 121, 139
Hydrogeological mapping, 201, 212, 229

I

Inertial measurement unit (IMU), 16
Inertial Navigation Systems (INS), 9
Inversed modeling algorithms, 79
investigation of glaciers, 234, 235, 243

L

land resource, 119, 120
low desert zone, 119

M

Machine learning (ML), 143, 144
magnetic profile, 168, 173, 178, 179, 181, 183
Maneuverability, 7
Manned airborne, 6
Maximum Chlorophyll Index (MCI), 76
Maximum Take-Off Weight (MTOW), 8
Medium Range Endurance, 8
Microwave Landing Systems (MLS), 11
multivariate, 144, 149, 151, 152, 155

B

P

Photogrammetry, 1, 6, 12, 15, 27, 28, 29, 31, 32, 34, 36, 41, 42, 46, 47, 49, 50, 51, 54, 57
Photogrammetry and Remote Sensing (ISPRS), 4
Photogrammetry and Remote Sensing (PaRS), 1
Phytoplankton, 59, 60, 61, 62, 64, 66, 67, 68, 69, 72, 73, 75, 76, 79, 80, 81, 83, 84, 85, 86, 87, 88, 89, 91, 92, 93, 94, 95, 96, 97, 98, 99, 100, 101, 102, 104, 105, 106, 107, 109, 110, 111, 112, 113, 114, 115, 116, 117
Phytoplankton bloom, 62

Q

quality control, 234, 235, 243
Quality-control tool, 12

R

Radiocommunication, 10
Rayleigh-corrected reflectance, 78
regional climate change, 233, 234
remote sensing, 201, 202, 229, 230, 231, 232
Remotely-Piloted Aerial System" (RPAS), 2
Roadmap, 14, 45, 48

S

satellite remote sensing, 233, 234, 235, 236, 239, 241, 242, 243, 244
self-organizing map, 143, 144
software application, 143
Spectral bands near, 73
Spectral difference, 74
structural element, 169, 171
sustainable development, 201, 202, 222, 230

T

Telecommunication, 10
Triggering, 60

W

Water temperature, 60
Worldwide Interoperability for Microwave Access (WiMAX), 11